Encyclopedia of Soybean: Biochemistry

Volume VIII

Encyclopedia of Soybean: Biochemistry

Volume VIII

Edited by **Albert Marinelli and Kiara Woods**

New York

Published by Callisto Reference,
106 Park Avenue, Suite 200,
New York, NY 10016, USA
www.callistoreference.com

Encyclopedia of Soybean: Biochemistry
Volume VIII
Edited by Albert Marinelli and Kiara Woods

International Standard Book Number: 978-1-63239-303-6 (Hardback)

Printed in the United States of America.

Contents

Preface

Every book is initially just a concept; it takes months of research and hard work to give it the final shape in which the readers receive it. In its early stages, this book also went through rigorous reviewing. The notable contributions made by experts from across the globe were first molded into patterned chapters and then arranged in a sensibly sequential manner to bring out the best results.

Soybean seed proteins are an economical and important source of protein in the diet of many developed and developing countries. As Soy is an efficient and major source of amino acids for human as well as animal nutrition. All the essential amino acids are provided by soybean protein in optimal amounts. It has been found through research that soy may also help in lowering the risks of colon, breast cancer and prostate. With the help of soy, osteoporosis and other bone health problems can also be avoided and hot flashes associated with menopause can be alleviated. The aim of this book is to serve as a good source of information about soybean for students as well as researchers. Some significant topics covered are Physiological Quality of Conventional Soybean Seeds, Mineral Nutrition, Asian Soybean Rust in South America and Soybean Cyst Nematode.

It has been my immense pleasure to be a part of this project and to contribute my years of learning in such a meaningful form. I would like to take this opportunity to thank all the people who have been associated with the completion of this book at any step.

Editor

Physiological Quality of Conventional and RR Soybean Seeds Associated with Lignin Content

Cristiane Fortes Gris[1] and Edila Vilela de Resende Von Pinho[2]
[1]Federal Institute of Southern Mines
[2]Federal University of Lavras
Brazil

1. Introduction

The sale of genetically modified soybean seed resistant to the Roundup Ready (RR) herbicide has revolutionized the worldwide soybean market in recent years. According to data from the International Service for the Acquisition of Agri-Biotech Applications-ISAAA (2009), in 2009, for the first time, more than three-quarters (77%) of the 90 million hectares of soybeans grown globally were biotech; followed by cotton, with almost half (49%) of the 33 million hectares being biotech; by maize, with over a quarter (26%) of the 158 million hectares grown globally being biotech; and finally by canola, with 21% of the 31 million hectares being biotech. These numbers indicate not only increases in hectares, but also a strong and growing adherence of farmers around the world to this technology.

Considering the area planted to RR soybeans in the 2009/10 growing season throughout the world, from these 69.3 million hectares, a demand of approximately 4.2 million tons of RR soybean seeds may be estimated, which makes the international soybean seed market ever more expressive and competitive. In Brazil alone, up to November 2010, nearly 35% of the total soybean cultivars registered in the Ministry of Agriculture were RR genetically modified, this number having increased more than 443% in the last four growing seasons, a result of the increase in the number of breeding programs for obtaining RR cultivars.

It is known that the physiological quality of soybean seeds is controlled in large part by the genotype or cultivar, features of the plant, and more specifically those of the pod and the seed itself, determining a differential response of each cultivar and its levels of tolerance to seed deterioration, to adverse field conditions and even to mechanized harvesting. Among seed characteristics, the seed coat is one of the principal conditioning factors for germination vigor and longevity of seeds, with its characteristics being associated with susceptibility to mechanical damage, longevity and potential for seed deterioration, which may be influenced by the lignin content and the degree of seed coat permeability. Understanding of the structure and properties of the seed coat has contributed to explaining and altering seed behavior under certain environmental conditions.

In the case of soybeans, differences in the lignin content among seed coat have been observed by various authors (Tavares et al., 1987; Carbonell et al., 1992; Alvarez, 1994; Carbonell & Krzyzanowski, 1995; Panobianco, 1997; Menezes, 2008). In addition, a great deal of speculation has been generated in relation to the lignin content in the plant between RR genetically modified soybean cultivars and conventional cultivars (Coghlan, 1999; Gertz

Junior et al., 1999; Kuiper et al., 2001; Edmisten et al., 2006; Nodari & Destro, 2006), indicating overproduction of this substance of up to 20% more in RR cultivars. Such variation may occur not only in the vegetative parts of plants, but also in reproductive parts, such as pods and seeds.

The term lignin is used to designate a group of substances with similar chemical units indicated as polymers derived from "p-coumaryl", "conyferyl" e "sinapyl" alcohols (Lewis & Yamamoto, 1990). Impermeable to water, lignin is also very resistant to pressure and not very elastic and it is the most abundant plant polymer after cellulose, being found in greater quantity in the cell wall, around 60% to 90% (Egg-Mendonça, 2001), and its deposition occurs during the formation of the cell wall.

According to the authors, overproduction of lignin observed in the RR soybean plant in the US, and more recently in Brazil, is leading to deep stem fissures, with a significant number of plants in the field presenting bent or broken stems, and this effect possibly arises in the presence of water deficit and high temperatures.

Although the exact cause of the lignin behavior in this mechanism is still unknown, the hypothesis of overproduction of lignin in RR soybean plants is based on the fact of the precursors of the lignin molecule being formed in the same metabolic pathway, the pathway of shikimic acid, inhibited by the glyphosate herbicide. The inhibition of EPSPS enzymes by glyphosate present in this pathway leads to a deficiency in the production of amino acids and consequent death of the plants. That way, the sequence CP4 EPSPS, introduced in the genome of commercial soybean cultivars responsible for production of the protein CP4 enolpyruvylshikimate-3-phosphate synthase (EPSPS), an enzyme that participates in the biosynthesis of aromatic amino acids in plants and microorganisms, may be presenting the pleiotropic effect, thus modifying the lignin content in the plant.

Nevertheless, research in this area is still quite limited and the few results published do not compare conventional cultivars with their respective RR genetically modified versions, but refer to comparison between diverse genotypes and therefore do not isolate the effect of the inserted transgene. In this context, it is relevant to discuss the results of more recent research dealing with this issue in this chapter, principally looking at comparisons between conventional materials and their RR versions, which are essentially derivatives.

For that reason, in this chapter we will discuss results of research dealing with the physiological quality and the lignin content in RR and conventional soybean seeds submitted to different harvest times and spraying with glyphosate herbicide, produced in two different time periods and submitted to direct imbibition in water.

2. The lignification process and RR soybeans

The term lignin is used to designate a group of substances with similar chemical units. According to Panobianco (1997), the chemical structure of lignin is very complex and still not very well defined. Butler & Bailey (1973), cited by Silva (1981), refer to lignin as a polymer, 3-methoxy-phenyl-propanol and 3-5-dimethoxy-phenyl-propanol, bonded in varied proportions and in random sequence, leading to a great variety of products, which makes exact definition difficult. According to Esau (1976), lignin consists of an organic substance or mixture of organic substances with high carbon content, but different from carbohydrates, and which is found associated with cellulose on the walls of numerous cells.

The term lignin is used to designate a group of substances with similar chemical units reported as polymers derived from "p-coumaryl", "conyferyl" e "sinapyl" alcohols (Lewis &

Yamamoto, 1990). Impermeable to water, lignin is also very resistant to pressure and not very elastic and it is the most abundant plant polymer after cellulose, being found in greater quantity in the cell wall, around 60% to 90% (Egg-Mendonça, 2001), and its deposition occurs during the formation of the cell wall.

The growth and development of the cell wall may be divided into two phases: growth of the primary wall, a phase in which the cell increases in size, and growth of the secondary wall, a phase in which deposition of lignin polymers occurs to the extent that the cell wall becomes progressively thicker as of the internal edge of the primary wall, in the direction of the center of the cell. The inclusion of lignin on the cell wall originates in the middle lamella, going in the direction of the interior of the secondary wall. According to Jung & Alen (1995), the effect of this lignin deposition pattern makes the middle lamella/primary cell wall region more intensely lignified.

This lignin deposition is important not only to lend rigidity and resistance to plant tissue, such as stem and leaves, but especially for the seed coat of soybean seeds, it has been correlated with resistance to mechanical damage (Alvarez, 1994; Panobianco, 1999), providing mechanical resistance to the tissue and protection against infestations by microorganisms to the cell wall (Rijo & Vasconcelos, 1983, cited by Tavares et al., 1987).

2.1 Lignification and the soybean seed coat

The seed coat is one of the main factors which determine germination capacity, vigor and longevity of seeds. It has a protective function during imbibition, avoiding cell rupture and loss of intracellular substances (Duke & Kakefuda, 1981), and also protects the embryonic axis (Carvalho & Nakagawa, 2000). It is derived from the integuments of the ovule where the primine gives rise to the testa and the secundine gives rise to the tegmen.

By means of a cross section of the testa of a soybean seed, three layers may be distinguished, the epidermis, the hypodermis and the inner parenchyma (Swanson et al., 1985). This last layer, composed of the spongy parenchyma, is present in the entire testa of the seed, except for the hilar region. It has from 6 to 8 cell layers, tangential to the surface of the testa, formed by thin walls and absent protoplasm, with the outermost part of this parenchyma being formed by large, elongated cells, while the innermost part by smaller and significantly branched cells (Esau, 1977).

The intermediate layer of the testa, the hypodermis, is formed of cells in hourglass form, or pillar cells, or even osteosclereid cells. It consists of a uniform cell layer through the entire testa, except for the hilar region. The cell wall of its sclerenchyma cells is not uniform, with the presence of large intercellular spaces (Corner, 1951).

The epidermis, outside of the testa, remains uniseriate and gives rise to the palisade layer, characteristic of leguminosae seeds. This layer consists of macroscereids (Malpighi cells) with wall of unequal thickness, having a cuticle present over their outermost wall. It cells are elongated and arranged perpendicular to the surface of the testa, with thick cell walls (Esau, 1976).

In soybean seeds, the thickness of the four testa layers altogether, including the cuticle, starting from the surface, may vary from 70 to 100 micrometers, there being variation among cultivars. Nevertheless, this characteristic is a constant with each cultivar and is controlled genetically (Caviness & Simpson, 1974). The presence or lack of pores and their quantity, shape and size on the surface of the testa is also controlled genetically. The pores seem to be related to water absorption, such that in hard seeds they are either absent or they exist in small quantity (Calero, 1981).

Morphological characteristics associated with the thickness and structure of the seed coat has also been related to the quality of soybean seeds. With the aid of a Scanning Electron Microscopy (SEM), it is possible to obtain a direct image of the atoms on the surface of a material, formed by secondary electrons and emitted from the surface of the irradiated specimen by the beam of primary electrons or by those scattered, which, in spite of generating poorer quality images, may indicate differences in the elementary composition of the sample. Designed basically for surface examination of samples, SEM allows the observation of internal surfaces if fractured and exposed, using principally secondary electrons (Alves, 2006).

Silva (2003), by means of scanning micrography of transversal sections of the testa of soybean seeds of the cultivars M-Soy 8400 and M-Soy 8411 observed three visible cell layers: palisade cell layers, an hourglass cell layer, and spongy parenchyma cells. The author evaluated the behavior of these cell layers that compose the testa of soybean seeds when they were exposed to five periods of accelerated aging (0, 24, 48, 72 and 96 hours) at 42° C and approximately 100% relative air humidity. For the cultivars evaluated, reduction in the thickness of the testa of the soybean seed was verified, which suggests collapse of the cells that compose such layers, which may be related to reduction of germination potential.

Menezes et al. (2009) evaluating the thickness and structure of the soybean seed coat (Figures 1 and 2) and the association of these characteristics with the physiological quality of the seeds, concluded that traits used for evaluation of physiological quality may be correlated with the lignin content of the seed coat. Nevertheless, according to the author, it was not possible to establish a relationship between the physiological quality of the soybean seeds and the anatomical aspects of the seed coat evaluated by SEM, emphasizing the need for refining the methodologies available for this purpose due to the difficulties of establishing the work area of common structures on the seeds, and of having observed that cell structures vary in different positions on the seed coat, which makes comparison of these structures among seeds of different genotypes difficult. In spite of that, in a general way, it was possible to observe that the lignin thickness on the palisade cell layers was greater when compared to the hourglass cell layers.

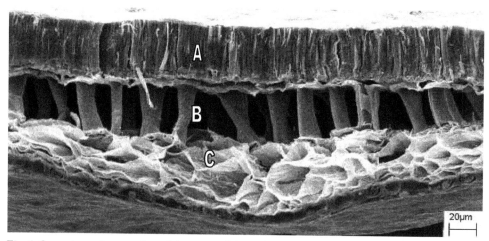

Fig. 1. Scanning micrography of the testa of the cultivar CD 201; A: palisade cell layer; B: hourglass cell layer and C: spongy parenchyma. Source: Menezes et al. (2009).

As is common in leguminosae, there is a particularly impermeable region on the walls of the upper part of the macrosclereids, which reflects light more intensely than the rest of the wall (Esau, 1965). What is called the conspicuous light line is visible in many wild soybean species, but is less prominent in cropped species (Alexandrova & Alexandrova, 1935, cited by Carlson & Lersten, 1987). This palisade layer drew the interest of researchers through the fact of its structure, and in certain hard seeds of leguminosae, being the cause of the high degree of impermeability of the seed coat, consequently affecting germination capacity (Esau, 1976).

Hard or impermeable seeds, according to Woodstock (1988), may be the result of compacted organization of cellulose microfibriles on the cell wall. This, for its part, may be impregnated with waterproof substances, such as lignin, waxes, suberins or tannin. They are abundantly composed of cellulose and hemicellulose polysaccharides, and of phenylpropanoid polymers such as lignin (McDougall et al., 1996).

In accordance with McDougall et al. (1996), the impermeability of the seed coat provided by lignin, exercises a significant effect on the speed and capacity of water absorption through it, thus interfering in the quantity of leached materials released to the outside during the imbibition phase of the seed germination process. Crocker (1948) already mentioned the need for better understanding of this mechanism since it was considered to be the best example of efficiency against water penetration and should therefore be better utilized by breeders in adjusting this characteristic to their needs. As general characteristics of soybean cultivars with a less permeable seed coat, one may cite better conservation potential, lower levels of infection by pathogens, greater vigor and viability, as well as resistance to reabsorption of moisture after maturation (Panobianco, 1999).

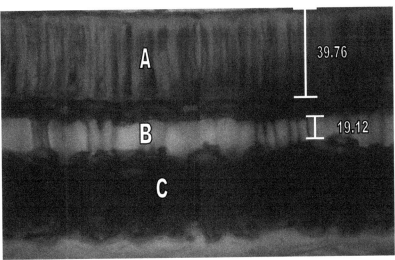

Fig. 2. Comparson of the thickness (µm) of the palisade cell and hourglass cell layers obtained by SEM from the cultivar CD 206. A: palisade cell layer; B: hourglass cell layer and C: spongy parenchyma. Source: Menezes et al. (2009).

Tavares et al. (1987), studying structural characteristics of the seed coat of seeds of soybean lines, concluded that the total fiber content is not connected with impermeability; however, in regard to the type of fiber, an accentuated increase in the lignin values was observed in the lines with impermeable seed coats (4.69% to 7.70%), differentiated from the values 1.80% to 3.18% found in lines with permeable seed coats. According to Brauns & Brauns (1960), cited by Tavares et al. (1987), the hydrophobic trait of lignin affects the hydrophilic bonds of the middle lamella and the removal of lignin interferes in the biological resistance of hydration in around 10.5% to 17% of the original tissue.

The occurrence of hard seeds in leguminosae has been attributed to both genetic and environmental factors (Donnelly, 1970). The percentage of hard seed exhibits considerable variability depending on the species or cultivar, the degree of maturity, the maturation conditions and the storage time. Thus, low air humidity during maturation results in a considerable increase in seed hardness (Baciu-Miclaus, 1970; Martins, 1989).

In soybeans, differences in the lignin content of the seed coat has been observed by various authors (Tavares et al., 1987; Carbonell et al., 1992; Alvarez, 1994; Carbonell & Krzyzanowski, 1995; Panobianco, 1999; Menezes et al., 2009; Gris et al., 2010;), and, in addition, differences have been reported in regard to the lignin content in the plant between genetically modified RR and conventional cultivars.

2.2 Lignin biosynthesis and RR soybeans

The advent of genetically modified soybeans, tolerant to the Roundup Ready© herbicide (RR), revolutionized the world soybean market. With the introduction of the CP4 EPSPS sequence in the genome of commercial soybean cultivars, which confers tolerance to the active ingredient glyphosate, the protein CP4 enolpyruvylshikimate-3-phosphate-synthase (EPSPS) is produced, an enzyme that participates in the biosynthesis of aromatic amino acids in plants and microorganisms. In the case of conventional cultivars, the inhibition of these enzymes by glyphosate, present in the shikimic acid pathway, leads to a deficiency in production of essential amino acids and consequent death of the plants, which does not occur in RR cultivars.

A great deal of speculation has been generated in relation to the lignin contents in the plant between genetically modified RR cultivars and conventional cultivars (Coghlan, 1999; Gertz Junior et al., 1999; Kuiper et al., 2001; Edmisten et al., 2006; Nodari & Destro, 2006).

In the late 1990s, some farmers in Georgia complained about the poor performance of their RR soybeans in years with a spring with drought and heat conditions. Scientists then carried out a comparative laboratory study of genetically modified and conventional soybeans (Gertz Junior et al. 1999). They found that the genetically modified plants were shorter, had a lower fresh weight, had less chlorophyll content, and, at high soil temperature of 40 °C to 50°C, suffered from stem splitting. According to Coghlan (1999), the elevated levels of lignin deposited in the stem of soybean plants would be leading to this splitting due to the stiffening of the plants under high temperatures (45°C), a problem also detected in genetically modified RR soybean crops in the USA, and which was to have led to considerable losses through falling of plants in hotter years (Nodari & Destro, 2006) as a consequence of overproduction of lignin in RR cultivars (Kuiper et al., 2001).

According to these authors, under stress conditions, losses in RR soybeans can arrive at 40% in comparison with conventional soybeans, brought about by greater production of lignin, up to 20% greater (Coghlan, 1999; Gertz et al., 1999). Nodari & Destro (2006), in a study undertaken in nine soybean crops in the state of Rio Grande do Sul (Brazil), observed that in the presence of drought and high temperatures, the RR soybean crops suffered more losses than conventional soybeans. The authors observed a large number of plants with deep stem splitting and a significant quantity of these plants had bent or broken stems, around 50% to 70% of the plants, according to the authors, possibly due to overproduction of lignin in the RR material (Figure 3).

Fig. 3. Plants of the "Maradona" variety with broken stem (left), split (middle) and intact stem (right). Source: Nodari & Destro (2006).

The plants are responsible for the production of secondary metabolites that perform innumerable functions, among which the terpenes, the phenolic compounds and the alkaloids are considered as the most important. The secondary compounds are biosynthesized through three basic metabolic pathways, the acetate-mevalonate, the acetate-malonate and the acetate-shikimate (Érsek & Kiraly, 1986), also denominated simply as mevalonic acid pathway, malonic acid pathway and shikimic acid pathway, respectively (Taiz & Zeiger, 1998).

In superior plants, the shikimic acid pathway occurs in plastids, there also being evidence that it is present in the cytosol (Hrazdina & Jensen, 1992). This important metabolic pathway begins with phosphoenolpyruvate (PEP), derived from glycolysis, and the erythrose 4-P coming from the monophosphate pentose pathway and the Calvin cycle, resulting in the biosynthesis of the phenylalanine amino acids, tyrosine and tryptophan (Salisbury & Ross, 1992) (Figure 4).

According to Resende et al. (2003), the enzymes that participate in the initial and intermediary steps of the lignin biosynthesis pathway are common to the phenylpropanoid pathway (Figure 5). The metabolism of the phenylpropanoids includes a complex series of biochemical pathways that provide the plants with thousands of combinations. Many of these, according to Boatright et al. (2004), are intermediate in the synthesis of structural substances of the cells, such as lignin, if formed from shikimic acid, which forms the basic units of the cinnamic and p-coumaric acids (Simões & Spitzer, 2004).

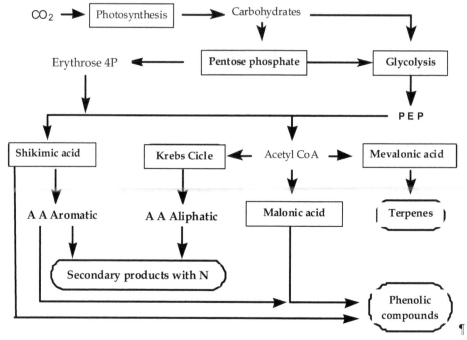

Source: Adapted from Taiz & Zeiger (1998).

Fig. 4. Schematic representation of the shikimic, malonic and mevalonic acid pathways. ¶

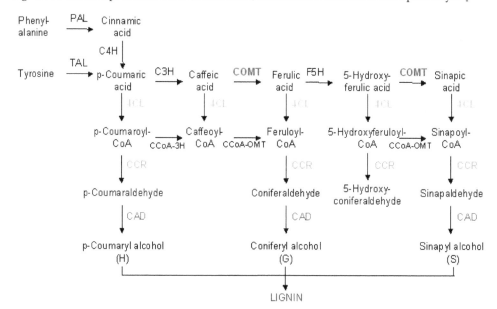

Fig. 5. Lignin biosynthesis pathway. Source: Baldoni (2010).

Lignin synthesis involves various enzymes and knowledge of them is important in studies in which the quality of soybean seeds and the lignin content is related (Baldoni, 2010). The complexity of the lignin biosynthesis pathways is attributed to various multifunctional enzymes, which also correspond to different gene families (Xu et al., 2009).

A considerable quantity of genes is attributed as participant in lignin synthesis, such as genes which regulate the activity of the enzymes phenylalanine ammonia-lyase (PAL), Cinnamate 4-Hydroxylase (C4H), 4-cumarate-CoA ligase (4CL), 4 Hydroxycinnamate 3-Hydroxylase (C3H), 5-Adenosine-Methionine: Caffeate/5-Hydroxy (OMT), Ferulate-5-Hydroxylase (F5H), Hydroxycinnamoyl COA Reductase (CCR), cinnamyl alcohol dehydrogenase (CAD) (Boudet, 2000; Boudet, 2003; Darley et al., 2001).

Although the exact cause of lignin behavior under stress conditions in RR soybean cultivars is still unknown (Coghlan, 1999), possibly the alterations in the content of this biopolymer in the plant is due to the fact of the precursors of the lignin molecule being formed in the shikimic acid pathway, which is inhibited by the glysophate herbicide in conventional plants. The inhibition of EPSPS enzymes, present in this pathway by the glyphosate, lead to a deficiency in the production of amino acids and consequent death of the plants. That way, the CP4 EPSPS sequence introduced in the genome of the commercial soybean cultivars denominated RR, responsible for the production of the protein CP4 enolpyruvylshikimate-3-phosphate synthase (EPSPS), an enzyme that participates in the biosynthesis of aromatic amino acids in plants and microorganisms, may present the pleiotropic effect, thus modifying the lignin content in the plant.

In spite of all those studies suggesting the pleiotropic effect of the transgene under high stress conditions in laboratory tests in the USA, some authors suggest that it might not be detected until specific environmental conditions are observed, which usually does not occur in field conditions. In this sense, the quantification of lignin in the plant, and consequently in pods and the seed coat of soybeans, become necessary in field conditions, principally with a view toward comparisons between conventional materials and their RR versions, which are essentially derivatives, since the previous reports refer to diverse genotypes, thus not isolating the effect of the inserted transgene. It is worth highlighting that scientific studies that truly prove the pleiotropic effect of the RR transgene under any characteristics are rare in the literature, with most of them being based only on observations and not on scientific results.

Therefore, we will further discuss some results of research obtained in Brazil in which the relation lignin versus RR and conventional soybean cultivars under diverse aspects was evaluated, emphasizing contents of this polymer in the plant, pod and seed coat.

3. Conventional and RR genetically modified soybeans: Some results in Brazil

3.1 Physiological quality and lignin content in the seed coats submitted to different harvest times

The viability period of the soybean seed is extremely variable, depending both on genetic characteristics and environmental effects during the phases of development, harvest, processing and storage. Once unfavorable conditions occur in some of these phases, physiological damages may result in losses to seed quality, with the intensity of these damages varying with the genetic factors intrinsic to each cultivar. Various researchers have emphasized the possibility of use of the seed with seed coat with a certain degree of impermeability to water as an alternative for avoiding loss of quality in the field (Gilioli & França Neto, 1982; Peske & Pereira, 1983; Hartwig & Potts, 1987), with delay in harvest and determination of the lignin content in the seed coat being methodologies suggested for genetic breeding programs for evaluation of the quality of soybean seeds (França Neto & Krzyzanowski, 2003).

Within this context, the work presented below (Gris et al., 2010) was conducted with the purpose of evaluating the physiological quality and lignin content in the seed coat of the conventional and RR soybean seeds collected at three different times in Lavras (MG), Brazil. Thus, the seeds of ten cultivars collected at stages R7, R8 and 20 days of harvest delay (R8+20) were submitted to tests for evaluation of physiological quality and lignin content. Harvest stages were determined according to Fehr & Caviness (1977).

We observed differences in the physiological quality of seeds among the different harvest times for the cultivars BRS 134, BRS 247 RR, Conquista, Jataí and Silvânia RR, with reduction in viability with harvest delay (R8 + 20). In a similar way, when submitted to accelerated aging, the seeds of the cultivars BRS 245 RR, BRS 134, BRS Jataí and Silvânia RR also underwent a reduction in vigor with harvest delay (Table 1). Braccini et al. (2003), studying the response of 15 genotypes of soybeans to harvest delay, also observed a significant reduction in germination percentage and vigor of seeds when they were submitted to harvest 30 days after the R8 stage of development.

Cultivars	Germination			Accelerated Aging			Electrical Conductivity		
	R7	R8	R8 + 20	R7	R8	R8 + 20	R7	R8	R8 + 20
Celeste	94.75a	96.50a	95.50a	94.75a	97.50a	91.50a	77.01a	82.42a	94.76a
Baliza RR	94.25a	93.00a	91.00a	91.50a	88.50a	84.00a	83.47b	90.86b	118.01a
BRS 133	91.25a	93.00a	88.00a	91.25a	96.50a	87.50a	93.66b	82.61b	107.15a
BRS 245 RR	91.75a	96.50a	90.50a	97.75a	99.50a	87.50b	94.79a	99.11a	97.37a
BRS 134	91.75a	90.50a	79.00b	94.00a	95.00a	75.50b	86.81a	87.02a	97.65a
BRS 247 RR	96.50a	98.50a	87.50b	94.00a	96.00a	87.00a	76.38a	85.18b	102.44b
Conquista	85.75a	90.00a	78.00b	89.75a	88.50a	84.00a	93.87b	85.23b	118.25a
Valiosa RR	89.75a	83.00a	84.50a	87.50a	92.00a	87.50a	98.15b	90.01b	112.56a
Jataí	91.50a	89.00a	76.50b	93.25a	87.00a	64.00b	83.42b	88.43b	152.70a
Silvânia RR	93.00a	91.50a	82.00b	92.50a	92.00a	71.00b	92.92b	89.61b	143.74a

Means followed by the same letter in the line for each determination do not differ among themselves by the Scott-Knott test at the 5% significance level.

Table 1. Means of the germination and accelerated aging test (% of normal seedlings) from seeds of soybean cultivars and their respective RR genetically modified forms, 2007/08 harvest. UFLA, Lavras, MG, Brazil.

We observed that the greatest decreases in seed vigor by the accelerated aging test (Table 1), when the harvest delay and the mean of stages R7 and R8 are contrasted, occurred for the cultivars Jataí and Silvânia RR, which presented, on average, losses in vigor of 40.82% and 29.93% respectively, indicating that not always cultivars that have high seed quality when collected near physiological maturity have greater tolerance to deterioration with delay of harvest. And, moreover, the greatest values of electrical conductivity were observed for the majority of seeds of the cultivars collected 20 days after the R8 stage, with exception of the cultivar BRS 247 RR, in which reduction in seed vigor was observed as of the R8 stage, and of the cultivars Celeste, BRS 245 RR and BRS 134 that did not undergo any alterations with the time of harvest.

As degradation of the cellular membranes is constituted hypothetically in the first event of the deterioration process (Delouche & Baskin, 1973), tests that evaluate membrane integrity, such as the electrical conductivity test, would theoretically be the most sensitive for estimating seed vigor, which is in agreement with the results obtained in this study, in which said test stood out in detecting differences of viability between the harvest times in seven of the ten cultivars evaluated. We emphasize that the electrical conductivity values observed in this study were situated from 77.01 μS cm^{-1} g^{-1} to 98.15 μS cm^{-1} g^{-1} for the R7 harvest time, 82.42 μS.cm^{-1}.g^{-1} to 99.11 μS.cm^{-1}.g^{-1} for the R8 harvest time and 94.76 μS.cm^{-1}.g^{-1} and 152.70 μS.cm^{-1}.g^{-1} for the 20 days after R8, values which demonstrate the growing trend of leachates released by the seeds with delay in harvest.

When we analyze the percentage of mechanical damage in seeds (Table 2), we observe the greatest values with delay of harvest for the cultivars Conquista (12.5%), Jataí (16.0%) and Silvânia RR (15.0%), which was not observed for the other cultivars studied. In addition, we also observed that by the germination test of seeds submitted to the water immersion test, three of the ten cultivars evaluated were differentiated in regard to the percentage of normal seedlings, however, with distinct responses. The lowest germination values when collected in R8 were observed in seeds of the cultivar BRS 245 RR; in those of the cultivar BRS 247 RR there was a reduction in germination when collected in R8 and R8 + 20; and finally in those of the cultivar Silvânia RR the lowest germinative power was verified when collected in R7 and R8. Various authors emphasize that soybean cultivars and lines behave differently in regard to degree of tolerance to delay of harvest (Lin & Severo, 1982; Rocha, 1982; Boldt, 1984), indicating that this trait may influence maintenance of the physiological quality of the seeds.

For the lignin content in the soybean seed coat, we can observe greater lignin content in the seed coat of seeds collected in the R7 and R8 + 20 stages, as well as for the cultivar Silvânia RR, when contrasted with its conventional version Jataí (Table 3).

When we observe the data of percentage of deformed abnormal seedlings, characterized by root curling, typical of damage by rapid imbibition, we observe a smaller number of abnormal seedlings due to the greater number of dead seeds with harvest delay. Giurizatto et al. (2003) affirm that the deteriorated seeds imbibe more rapidly and are therefore more prone to greater damage through imbibition, which is in agreement with the results obtained in this study.

According to Alpert & Oliver (2002) the cellular membranes have two main states, one more fluid or "crystalline liquid" and another less fluid or "gel", remaining, when organized, in the crystalline phase. In a dry seed, the membranes are found in the gel phase and therefore do not constitute an efficient barrier to contain the release of solutes. When the seeds are exposed to rapid imbibitions, the water penetrates before the membrane can be reverted to the crystalline liquid phase, with damage occurring to the cells; thus, the transition between these two phases in the configuration of the membrane constitutes the fundamental cause of possible injuries during imbibition of seeds, which makes the study of the role of lignin in the seed coat even more important.

Cultivars	Mechanical Damage			Germination after Immersion Normal seedlings			Germination after Immersion Abnormal curled seedlings		
	R7	R8	R8 + 20	R7	R8	R8 + 20	R7	R8	R8 + 20
Celeste	3.50a	2.50a	3.00a	62.50a	70.50a	62.00a	20.00a	14.50a	11.00a
Baliza RR	3.00a	3.00a	6.00a	50.00a	46.50a	44.50a	17.50a	24.00a	12.00a
BRS 133	3.00a	1.00a	2.00a	55.00a	49.50a	43.50a	14.50b	26.00a	17.00b
BRS 245 RR	2.50a	2.50a	5.00a	46.00a	22.50b	43.50a	19.00a	17.00b	32.00b
BRS 134	1.50a	1.50a	1.00a	51.00a	47.50a	36.00a	26.00a	26.50a	23.00a
BRS 247 RR	1.50a	1.00a	3.50a	63.00a	50.50a	41.00b	15.50b	32.00a	23.00b
Conquista	6.00b	4.50b	12.50a	38.00a	33.00a	35.00a	10.50a	6.00a	1.00a
Valiosa RR	5.50a	4.50a	5.50a	35.50a	25.50a	36.00a	9.00a	4.50a	3.50a
Jataí	2.50b	3.50b	16.00a	20.50a	29.50a	26.00a	32.00a	41.50a	2.50b
Silvânia RR	4.50b	5.00b	15.00a	21.50b	28.50b	40.50a	30.00a	26.00a	1.50b

Means followed by the same letter in the line for each determination do not differ among themselves by the Scott-Knott test at the 5% significance level.

Table 2. Means obtained for mechanical damage (%) and germination after water immersion (% of normal seedlings and abnormal curled) of soybean cultivar seeds and their genetically modified RR forms, 2007/08 harvest. UFLA, Lavras, MG, Brazil.

These differences observed for the lignin content among the harvest times are not biologically explainable, having possibly been detected due to the low coefficient of variation (CV) obtained for this variable. When we analyze the sole significant contrast, for its part, the genetically modified cultivar Silvânia RR presented greater lignin content in the seed coat than its respective conventional cultivar Jataí. Nevertheless, as an isolated fact, among the five RR combinations versus the conventional versions tested, in our view it does not justify a greater inference regarding pleiotropy of the RR transgene.

Lignin Content				
Harvest Stages			Cultivars	
R7	R8	R8 + 20	Jataí	Silvânia RR
0.2685a	0.2385b	0.2615a	0.3008b	0.4167a

Means followed by the same letter in the column do not differ among themselves by the Scott-Knott test at the 5% significance level.

Table 3. Means with a significant difference obtained for lignin content in the soybean seed coat (%), 2007/08 harvest, UFLA, Lavras, MG, Brazil.

In general, we can conclude that in spite of there being behavioral differences in regard to tolerance to harvest delay among the different cultivars evaluated, we did not observe consistent results in regard to a comparison of the RR versus conventional cultivar, not indicating, for the conditions of this test, any sign of pleiotropy.

3.2 Physiological quality and lignin content in the plants submitted to spraying with glyphosate

Glyphosate (N-phosphonomethyl glycine) is one of the most used herbicides in weed control throughout the world, making up nearly 12% of global herbicide sales and presenting more than 150 commercial brands (Kruse et al., 2000). The emergence of RR genetically modified soybeans increased the use of this molecule in soybeans crops in a considerable way and, along with this, also the environmental concern due to exclusive and indiscriminate use of this herbicide.

According to Sanino et al. (1999), although pesticides (especially glyphosate) may have a beneficial effect on agricultural productivity, the potential risk of these chemical compounds in the environment must be considered, which makes greater studies regarding the behavior of glyphosate under tropical conditions relevant. Within this context we aimed to evaluate the physiological quality of genetically modified RR soybean seeds and the lignin contents of plants submitted to spraying with glyphosate herbicide (Gris, 2009).

In Tables 4 and 5 we present the mean results for the variables analyzed when the soybean plants were submitted to spraying with glyphosate herbicide and water (greenhouse test) and spraying with glyphosate herbicide or manual weeding (field test) respectively.

Cultivars	Germination		Accelerated Aging		Seed Coat Lignin	
	Water	Herbicide	Water	Herbicide	Water	Herbicide
Valiosa RR	91.0a	89.50a	98.75a	97.50a	0.33a	0.26a
BRS 245 RR	94.25a	93.75a	98.50a	96.75a	0.21a	0.22a

Means followed by the same letter in the line for each determination do not differ among themselves by the Scott-Knott test at the 5% significance level.

Table 4. Means of germination and Accelerated aging (% of normal seedlings) and lignin content in the seed coat (%) of genetically modified RR soybean seeds submitted to spraying with water and glyphosate herbicide, 2007/08 harvest, Lavras, MG, Brazil, greenhouse test.

We observe that application of the glyphosate herbicide did not alter the physiological quality of the soybean seeds nor the lignin contents in the seed coat and in the plant for the two tests evaluated. These results are not in agreement with those obtained by Sanino et al. (1999), who studying the effect of application of glyphosate herbicide in soybeans observed, in a general way, reduction in the physiological quality of the RR seeds, as well as considerable reduction in activity of the enzyme α-amylase in terms of time. It is worth emphasizing that such a study was carried out comparing only 2 soybean cultivars, one conventional and one genetically modified RR variety, and that the two did not represent the same genotype, since they originated from different parentages.

In this study (Gris, 2009) we obtained a significant response only for the interaction cultivar versus treatments, when the values of electrical conductivity of the seeds produced in the field test were evaluated (Table 5), in which we observed that seeds of the cultivars Baliza RR and BRS 247 RR had their values reduced and increased respectively when the same spraying was performed. Such a differential response may possibly be explained by the different capacity of the genes inserted in the RR cultivars in expressing tolerance to the glyphosate herbicide, which according to Lacerda & Matallo (2008) may or may not occur in a homogeneous manner among cultivars and even within the same cultivar, as well as other factors inherent to the genetics of each cultivar.

Cultivars	Germination		Accelerated Aging		Mechanical Damage		ESI	
	Weeding	Herbicide	Weeding	Herbicide	Weeding	Herbicide	Weeding	Herbicide
Baliza RR	93.50a	96.50a	88.00a	90.50a	0.75a	0.75a	7.07a	7.14a
BRS 245 RR	89.00a	93.50a	97.00a	95.50a	1.00a	0.00a	7.24a	7.12a
BRS 247 RR	96.50a	97.75a	94.50a	91.50a	0.75a	1.75a	7.19a	7.07a
Silvânia RR	93.00a	93.00a	87.00a	88.00a	3.00a	2.00a	7.44a	7.00a
Valiosa RR	86.00a	89.00a	84.50a	83.50a	2.00a	2.25a	7.42a	7.39a
Cultivars	Electrical Conduct.		Seed Coat Lignin		Pod Lignin		Stem Lignin	
	Weeding	Herbicide	Weeding	Herbicide	Weeding	Herbicide	Weeding	Herbicide
Baliza RR	61.0a	48.0b	0.24a	0.23a	8.66a	8.09a	12.71a	13.61a
BRS 245 RR	69.0a	70.0a	0.19a	0.19a	8.41a	7.81a	12.20a	13.07a
BRS 247 RR	52.0b	70.0a	0.20a	0.20a	9.26a	8.57a	13.43a	12.91a
Silvânia RR	69.0a	69.0a	0.29a	0.30a	9.61a	9.03a	20.23a	18.50a
Valiosa RR	46.0a	40.0a	0.27a	0.30a	7.87a	7.70a	14.62a	13.35a

Means followed by the same letter in the line for each determination do not differ among themselves by the Scott-Knott test at the 5% significance level.

Table 5. Means of germination and Accelerated aging (% of normal seedlings), Mechanical damage (%), Emergence speed index – ESI (days), Electrical conductivity ($\mu S.cm^{-1}.g^{-1}$), Lignin content in the seed coat, pod and stem (%) of genetically modified RR soybean cultivars submitted to manual weeding and spraying with glyphosate herbicide, 2007/08 harvest, Lavras, MG, Brazil, field test.

It is worth emphasizing that since degradation of the cellular membranes is constituted hypothetically in the first event of the deterioration process (Delouche & Baskin, 1973), tests such as electrical conductivity that evaluate membrane integrity are theoretically most sensitive for estimating seed vigor, which possibly, allied with the affirmations of Lacerda & Matallo (2008), would explain the alterations only in the conductivity values.

The absence of a significant response for treatments with weeding and spraying with the glyphosate herbicide indicate that in a general way they did not influence the physiological quality of the seeds, nor the lignin content in the soybean plants. According to Cole & Cerdeira (1982) the blocking of the shikimate pathway due to the action of the glyphosate leads to the accumulation of shikimic acid with many physiological and ecological implications, which, according to Duke & Hoagland (1985) and Becerril et al. (1989), may result in synthesis of indol acetic acid of other plant hormones, chlorophyll synthesis, phytoalexin and lignin synthesis and protein synthesis, and affect photosynthesis, respiration, transpiration, permeability of membranes and other factors.

In addition, other studies have shown that applications of glyphosate in crops interfere in nutrient absorption, increase pests and diseases, reducing crop vigor and yield (Antoniou et al., 2010). According to compilation of data made by these authors, glyphosate reduces nutrient absorption by plants, immobilizing trace elements such as iron and manganese in the soil, as well as avoiding their transport from the roots to the above ground part

(Strautman, 2007). As a result, RR soybean plants treated with glyphosate have lower levels of manganese and other nutrients and reduction in growth of budding and roots (Zobiole et al., 2010). It is worth emphasizing that the seeds produced in the two tests described in this secondary heading are being tested in regard to variation in chemical composition, data which should soon be published.

Both in the field test and in the greenhouse test, it was not possible to relate physiological quality of the seeds and lignin content in their seed coat. We observed significant differences only among the cultivars evaluated, which presented different responses when submitted to the different vigor tests, as well as lignin content, which was already expected, in terms of the great genetic variability among them.

We conclude from these tests that there is a differential response for the electrical conductivity values of the seeds when the plants of different soybean cultivars are submitted to spraying with the glyphosate herbicide; nevertheless, we did not observe a difference in the lignin contents in the stem, in the pod and in the seed coat of the soybean seeds in the cultivars evaluated when submitted to spraying with the glyphosate herbicide.

3.3 Agronomic characteristics and quality of soybean seeds produced at different times

It is known that different planting times, influenced by different environmental conditions, may be determining factors for the development of seed deterioration tolerance mechanisms and therefore for the quality of soybean seeds. Considered as a seed deterioration tolerance mechanism, the impermeability of the seed coat, characterized principally by seeds with greater lignin content, hinders water penetration in the seed coat. In a similar way to alterations in the germination process and in manifestation of vigor, in terms of the climate in the seed production phase, environmental conditions may also in some way affect the metabolism and chemical constitution of the seeds.

As we have already seen in this chapter, according to some authors, overproduction of lignin in RR soybean plants may be associated with the presence of water deficit and high temperatures during cropping, indicating that the environmental conditions found in the field during crop development may affect lignin production in the plant in an expressive way.

With this objective, we compared agronomic traits of the plant, physiological quality and seed health and lignin content in the seed coat of RR and conventional seeds produced in different time periods, summer and winter (Gris, 2009), with the determinations: plant height, height of insertion of the first pod and number of pods per plant, weight of 1000 seeds (Brasil, 1992), lignin content in the seed coat (Capeleti et al., 2005), incidence of mechanical damage (Marcos Filho et al., 1987), germination and dry matter of normal seedling from germination (Brasil, 1992), emergence speed index and germination speed index (Edmond & Drapala, 1958), final stand in the seed bed (counting at 24 days after seeding), accelerated aging at 42°C for 72h (Marcos Filho, 1999), electrical conductivity (Vieira, 1994), water immersion test of seeds and seed health, evaluating the infestation percentage (Machado, 2000) and intensity of the inoculums. The data of inoculum density were weighted by the McKinney formula (1923):

$$II(\%)=\frac{\Sigma(F \times n) \times 100}{(N \times M)}$$

In which: II = inoculum intensity, F = number of seeds with a determined score, n = score observed, N = total number of seeds evaluated and M = maximum score of the scale.

In Table 6, we present a summary of the mean results for the variables in which the contrasts (RR cultivar versus conventional cultivar) presented a significant difference, for both harvests, in which among all the characteristics evaluated, significant results for the contrasts evaluated were few.

For the electrical conductivity test, we observed a greater value for the conventional cultivar Jataí (76.54 $\mu S.cm^{-1}.g^{-1}$) when compared to the cultivar Silvânia RR (100.25 $\mu S.cm^{-1}.g^{-1}$). According to Vieira & Krzyzanowski (1999) for lots of high vigor soybean seeds, the standard conductivity values should be situated at most up to 70-80 $\mu S.cm^{-1}.g^{-1}$, however with a strong trend to present medium vigor. Nevertheless, in spite of the high value of electrical conductivity observed in seeds of the cultivar Silvânia RR, we did not observe differences between the two cultivars in the germination and vigor tests, which, according to José et al. (2004), may indicate that there are cultivars with greater efficiency in membrane reorganization, not resulting in damages, strictly speaking.

Variables		Means – Summer 2006/07 harvest			
Plant height (m)	Jataí	1.56 a	vs	Silvânia RR	1.41 b
Number Pods/plant	Jataí	110.00 a	vs	Silvânia RR	57.50 b
Germination (%)	BRS 133	95.50 a	vs	BRS 245 RR	87.25 b
Weight of 1000 seeds (g)	BRS 134	155.50 a	vs	BRS 247 RR	142.70 b
Emergence Speed Index	BRS 134	7.16 b	vs	BRS 247 RR	7.55 a
Lignin Seed Coat (%)	Celeste	0.20 b	vs	Baliza RR	0.26 a
Variables		Means – Winter 2007 harvest			
Electrical conductivity ($\mu S.cm^{-1}.g^{-1}$)	Jataí	76.54 b	vs	Silvânia RR	100.25 a

Capital letters followed by the same letter in the line do not differ among themselves by the Scheffe Test, at the 5% significance level.

Table 6. Mean values for some variables in which the contrasts between the conventional soybean cultivar and its genetically modified RR version presented significance, summer and winter harvest, Lavras, MG, Brazil.

Panobianco (1997) upon reporting variation in electrical conductivity of soybean seeds and the lignin content in their seed coat affirms that the genotype may alter the electrical conductivity for seeds with the same standard of physiological quality. Nevertheless, we did not observe significant differences between the cultivars Jataí and Silvânia RR in regard to lignin content in the seed coat, indicating that, in this case, it may not have been responsible for the variation in electrical conductivity observed. In the same way, it was not possible to relate the difference in the lignin contents in the seed coat, observed between the cultivars Celeste (0.20%) and Baliza RR (0.26%), and the results of physiological quality of the two, produced in the summer harvest, since they differed only for this characteristic. It is worth highlighting that in spite of the differences found for these two cultivars, it was not possible through the incidence of mechanical damage to detect any differences between the cultivars studied.

Upon observing the contrasts established between the RR and conventional cultivars, we can infer that the cultivars Jataí and Silvânia RR presented the greatest number of significant differences among the variable studied (Table 6), not only in relation to the physiological quality of the seeds, but also in regard to agronomic traits, such as plant height and number of pods per plant.

When we analyze the mean values of plant height and number of pods per plant, we verify once more that the conventional cultivar Jataí showed superiority to the cultivar Silvânia RR, such that for number of pods/plant, these values were up to 91.3% greater. Nevertheless, it is worth emphasizing that for these two cultivars in field conditions, we observed the greatest variations in regard to the phenological cycle, with greater uniformity in maturation and a shorter cycle, around 10 days, of the conventional cultivar Jataí in relation to the genetically modified RR cultivar. It is fitting to highlight that in spite of the RR cultivars tested in this study being essentially derivatives of the respective conventional cultivars, by means of backcrossings, the genotype of the recurrent genitor is not always recovered, due to number fewer recurrence cycles which may consequently result in variations between both materials. Nevertheless, for these cultivars, there is no information on the number of backcrossing cycles used.

When we evaluate the physiological quality of the seeds by means of the germination test in the summer harvest and of the germination speed index (IGV) in the winter harvest, we do not observe a relationship between the significant results for these variables, with the contrasts BRS 133 versus BRS 245 RR and Conquista versus Valiosa RR being differentiated respectively. For both results, the conventional cultivars showed superiority to the genetically modified RR cultivars, with the conventional cultivar BRS 133, with 95% of normal seedlings, overcoming the cultivar BRS 245 RR, with 87%, by approximately 9.5%, when they were produced in the summer harvest. Nevertheless, by the results in reference to the Emergence Speed Index, we observe a lower value for the genetically modified cultivar BRS 247 RR (7.55 days) in comparison with the conventional cultivar BRS 134 (7.16), which once more shows the inconsistency of data that justify a pleiotropic effect of the RR gene on lignin production.

It is worth emphasizing that in spite of the results found in this study, with exception of the variables Emergence Speed Index and lignin in the seed coat, the RR cultivars stood out in relation to the conventional cultivars; most of the significant contrasts, were seen to be isolated, in only one of the harvests or one of the tests in the midst of various comparisons among physiological quality of the seeds, therefore not indicating substantial differences of quality between the RR and conventional materials.

According to Menezes (2008) the physiological quality of soybean seeds is influenced by the maternal or extra-chromosome effect, just as is the cytoplasmatic inheritance, with the physical characteristics of the seed coat, of maternal origin, not being sole determinants of the physiological quality of the seeds. According to this author, the study of genetic control for seed quality indicates the effect of the general and specific combination capacity, which suggests the presence of additive and non-additive gene effects for physiological quality of soybean seeds. Therefore, the quality of seeds may not be attributed only to their seed coat and consequently to their lignin contents, but also to genes present in the nucleus.

When we analyze the results obtained in the seed health test (Figure 6), we observe that the cultivars BRS 133, BRS 245 RR, BRS 134 and BRS 247 RR presented the lowest percentages of infection and infection indexes (severity), when produced in the summer, indicating that the environmental conditions during the seed maturation period were responsible for seed health quality. In these cultivars a shorter phenological cycle and semi-early maturity was observed, which provided for the maturation period outside of the rainy period.

According to Delouche (1975), the alternating of dry and wet days during the maturation phase until harvest, which occurs with greater facility in the summer, can increase the incidence of diseases in a differentiated way at the end of the cycle of the seeds produced. Within this context, the seed becomes not only an easy target for the action of microorganisms, which considerably reduce its viability, but they also come to be efficient vehicles for dissemination of pathogens (Machado, 2000). This situation may be visualized principally for the cultivars Jataí and Silvânia RR, which remained for a greater period in the field, and presented the greatest percentages of infection, 39% and 38% (Figure 6A), and also the greatest indexes of infection by the pathogen Phomopsis, 35% and 26% (Figure 6B), respectively. It is worth emphasizing that when produced in winter conditions, under a controlled irrigation system, without rains in the seed maturation period, the presence of pathogens was not observed for any seeds.

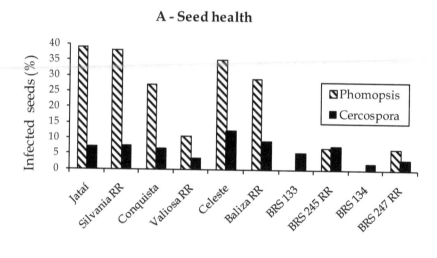

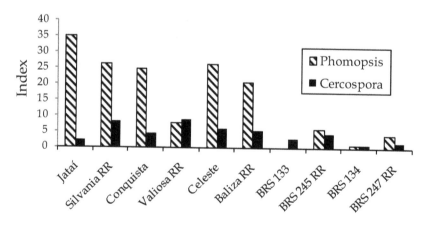

Fig. 6. Average values for infection percentage (A) and infection indexes - severity (B) in the seed health test of conventional soybeans and the genetically modified RR versions, summer harvest.

In relation to the RR versus conventional contrasts, we observe that in spite of the cultivars tested in this study having their origin in the same genotype by successive backcrossings, when observed in the field, we verified that some presented perceptible cycle variations, maintaining the cultivars Conquista and Celeste for more days in the field in relation to the cultivars Valiosa RR and Baliza RR, respectively; enough so that the first, subjected to rains and high temperatures, presented slightly greater values in the seed health and severity test. In this case, we cannot attribute the differences of RR versus conventional contrast, observed in Figure 6A and 6B, to the effect of the RR transgene, but rather to environmental conditions associated with difference of cycle.

In view of the above, in spite of some authors suggesting the pleiotropic effect of the transgene CP4 EPSPS on lignin overproduction in the plant, it was not possible for us to identify the pleiotropic effect in the cultivars studied in this and in the other studies described here, which indicates that the alterations of lignin content in the plant, observed by those authors under normal climatic conditions, are not due to the fact of the lignin molecule precursors being formed in the shikimic acid pathway. Thus, the sequence CP4 EPSPS, introduced in the genome of commercial soybean cultivars, responsible for the production of the protein CP4 enolpyruvylshikimate-3-phosphate-synthase (EPSPS), an enzyme that participates in the biosynthesis of aromatic amino acids in plants and microorganisms, seems not to be associated with lignin contents in the plant and in the soybean seed coat, and it seems that there are no substantial differences in regard to the agronomic traits and physiological quality of seeds between conventional and genetically modified RR cultivars.

4. References

Alpert, P.; Oliver, M.J. (2002). *Drying without dying*. In: BLACK, M.; PRITCHARD, H.W. (Ed.). Desiccation and survival in plants: drying without dying. Wallingford: CABI, 2002. p.4-43.

Alvarez, P. J. C. (1994). *Relação entre o conteúdo de lignina no tegumento da semente de soja e sua relação ao dano mecânico*. 43 p. Dissertação (Mestrado em Genética e Melhoramento) – Faculdade Estadual de Londrina, Londrina.

Alves, E. (2006). *Apostila do curso introdutório à microscopia eletrônica de varredura*. Lavras: UFLA, 43 p.

Antoniou, M.; Brack, P.; Carrasco, A.; Fagan, J.; Habib, M.; Kageyama, P.; Leifert, C.; Nodari, R.O.; Pengue, W. (2010). *Soja Transgênica:* Sustentável? Responsável? 2010. Available HTTP: http://www.gmwatch.org/files/GMsoy_Sust_Respons_FULL_POR_v2.pdf (11/20/2010).

Baciu-Miclaus, D. (1970). Contribuition to the study of hard seed and coat structure properties of soybean. *Proccedings of the International Seed Testing Association*, Vollebekk, v.35, n.2, p.599-617.

Baldoni, A. (2010). *Análises fisiológicas, ultraestruturais e expressão gênica de lignina em sementes de soja*. 64p. Thesis (Mestrado em Produção e Tecnologia de sementes) – Universidade Federal de Lavras, Lavras.

Becerril, J.M.; Duke, S.O.; Lydon, J. (1989). Glyphosate Effects on Shikimate Pathway Products in Leaves and Flowers of Velvetleaf. *Phytochemistry*, v.28, p.695-99.

Bervald, C.M.P. (2006). *Desempenho fisiológico e metabolismo de sementes de soja convencional e transgênica submetidas ao glifosato*. Pelotas. Dissertação (Mestrado). Universidade Federal de Pelotas.

Boatright, J. et al. (July 2004). Understanding in vivo benzenoid metabolism in petunia petal tissue. *Plant Physiology*, Bethesda, v. 135, n. 4, p. 1993-2011.

Boldt, A.F. (1984). *Relação entre os caracteres de qualidade da vagem e da semente de soja (Glycine max* (L.) Merrill). 70f. Dissertation (Mestrado em Fitotecnia)-Universidade Federal de Viçosa, Viçosa, MG.

Boudet, A. M. (Feb. 2000). Lignins and lignification: selected issues. *Plant Physiology and Biochemistry*, New Delhi, v. 38, n. 1/2, p. 81-96.

Boudet, A.M. (Aug. 2003). Lignins and lignocellulosics: a better control of synthesis for new and improved uses. *Trends in Plant Science*, Oxford, v. 12, n. 8, p. 576-581.

Braccini, A.deL. e; Albrecht, L.P.; Ávila, M.R.; Scapim, C.A.; Bio, F.E.I.; Pelegrinello, S.R. (Mar./abr. 2003). Qualidade fisiológica e sanitária das sementes de quinze cultivares de soja (Glycine max (L.) Merrill) colhidas na época normal e após o retardamento de colheita. *Acta Scientiarum Agronomy*, Maringá, v.25, n.2, p.449-457.

Brasil. (1992). Ministério da Agricultura. *Regras para análise de sementes*. Brasília: MA/SNDA/DNDV/CLV. 365p.

Calero, E.; West, S. H.; Hinson, K.(Nov./Dec. 1981). Water absorption of soybean seeds and associated causal factors. *Crop Science*, Madison, v. 21, n. 6, p. 926-933.

Capeleti, I.; Ferrarese, M.L.L.; Krzyzanowski, F.C.; Ferrarese Filho, O. (July 2005). A new procedure for quantification of lignin in soybean (*Glycine max* (L.) Merrill) seed coat and their relationship with the resistance to mechanical damage. *Seed Science and Technology*, Zurich, v.33, n.2, p.511-515.

Carbonell, S.A.M.; Krzyzanowski, F.C. (1995). The pendulum test for screening soybean genotypes for seeds resistance to mechanical damage. *Seed Science and Tecnology*, Zurich, v.23, n.2, p.331-339.

Carbonell, S.A.M.; Krzyzanowski, F.C.; Kaster, M. (Mar./abr. 1992). Avaliação do "teste de queda" para seleção de genótipos de soja com semente resistente ao dano mecânico. *Revista Brasileira de Sementes*, Brasília, v.14, n.2, p.215-219.

Carlson, J.B.; Lersten, N.R. (1987). *Reproductive morphology*. In: Wilcox, J.R. Soybeans: improvement, production and uses. Madison: ASA/CSSA/SSSA. p.95-134.

Carvalho, N. M.; Nakagawa, J. (2000). *Sementes:* ciência, tecnologia e produção. 4. ed. Jaboticabal: FUNEP, 588 p.

Caviness, C.E.; Simpsom, A.M.JR. (1974). Influence of variety and location on seed coat thickness of mature soybean seed. *Proceedings Association of seed Analanalysis*, Wellington, v. 64, p. 102-108.

Coghlan, A. (1999). *Splitting headache:* Monsanto's modified soya beans are cracking up in the heat. Saint Louis: Monsanto. Available at: <http://www.mindfully.org/GE/Monsanto-RR-Soy-Cracking.htm>. Accessed on: Mar. 10, 2009.

Cole, A.W.; Cerdeira, A.L. (1985). Southernpea response to glyphosate desiccation. *HortScience*, Alexandria, v.17, n.2, p.244-246, 1982.

Corner, E.J. (1951). The leguminous seeds. *Phytomorphology*, New Delhi, v.1, p.117-150.

Crocker, W. (1948). *Growth of plants*. New York: Reinohold. 459p.

Darley, C. P. et al. (Sept. 2001). The molecular basis of plant cell wall extension. *Plant Molecular Biology*, Dordrecht, v. 47, n. 1/2, p. 179-195.

Delouche, J.C.; Baskin, C.C. (1973). Accelerated aging techniques for predicting the relative storability of seed lots. *Seed Science and Technology*, Zurich, v.1, n.2, p.427-452.

Donelly, E.D. (1970). Persistance of hard seed in Vicia lines derived from interspecific hybridization. *Crop Science*, Madison, v.10, n.6, p.661-662.

Duke, S. H.; Kakefuda, G. (Mar. 1981). Role of the testa in preventing cellular rupture during imbibition of the legume seeds. *Plant Physiology*, Bethesda, v. 67, n. 2, p. 449-456.

Duke, S.O.; Hoagland, R.E. (1985). *Effects of glyphosate on metabolism of phenolic compounds.* Washington: CAB. Available at: <http://www.cababstractsplus.org/abstracts/Abstract.aspx?AcNo=19850776767>. Accessed on: Mar. 11, 2009.

Edmisten, K.L.; Wells, R.; Wilcut, J.W. (2000). *Investigation of the cavitation and large boll shed in roundup ready cotton.* Available at: <http://www.cottoninc.com/projectsummaries/2000ProjectSummaries/detail.asp?projectID=119>. Accessed on: Mar. 12, 2006.

Edmond, J.B.; Drapala, W.S. (June 1958). The effects of temperature, sand and acetone on germination of okra seed. *Proceedings of the American Society for Horticultural Science*, New York, v.71, p.428-434.

Egg-Mendonça, C.V. do C. (2001). *Caracterização química e enzimática de famílias de feijões obtidas do cruzamento das linhagens Amarelinho e CI – 107.* 48p. Dissertation (Mestrado em Agrobioquímica)-Universidade Federal de Lavras, Lavras.

Érzek, T. & Kiraly, Z. (1986). Phytoalexins and questions that remain unresolved forty-five years after their discovery. *Acta Phytopathologica et Entomolologica Hungarica*, 21: 5-14.

Esau, K. (1965). *Plant anatomy.* 2.ed. New York: J.Wiley. 767p.

Esau, K. (1976). *Anatomia das plantas com sementes.* São Paulo: E.Blucher. 293p.

Esau, K. (1977). *Anatomy of seeds plants.* New York: J.Wiley. 550p.

Fehr, W.R.; Caviness, C.E. (1977). *Stage of soybean development.* Ames: Iowa State University. 11p.

França Neto, J. de B.; Krzyzanowski, F.C. (2003). Estratégias do melhoramento para produção de sementes de soja no Brasil. In: SIMPÓSIO SOBRE ATUALIZAÇÃO EM GENÉTICA E MELHORAMENTO DE PLANTAS: Melhoramento de plantas e produção de sementes no Brasil, 7., 2003, Lavras. *Anais...* Lavras: UFLA, 2003. Available at: <http://www.nucleoestudo.ufla.br/gen/eventos/simposios/7simpo/resumos/20 036.pdf>. Accessed on: Apr. 22, 2006.

Gertz Junior, J.M.; Vencill, W.K.; Hill, N.S. (1999). Tolerance of transgenic soybean (Glycine mar) to heat stress. In: brighton crop protection conference: weeds, 3., Brighton. *Proceedings...* Brighton: BCP, 1999. p.835-840.

Gilioli, J.L.; França Neto, J.B. (1981). Efeito da escarificação mecânica e do retardamento de colheita sobre a emergência de sementes de soja com tegumento impermeável. In: SEMINÁRIO NACIONAL DE PESQUISA DE SOJA, 2., Brasília. *Anais...* Londrina: EMBRAPA-CNPSo, 1982. v.1, p.601-609. (EMBRAPA-CNPSo. Documentos, 1).

Giurizatto, M.I.K.; Ouza, L.C.F.; Robaina, A.D.; Gonçalves, M.C. (Jul./ago. 2003). Efeito da época de colheita e da espessura do tegumento sobre a viabilidade e o vigor de sementes de soja. *Ciência e Agrotecnologia*, Lavras, v.27, n.4, p.771-779.

Gris, C. F. (2009). Qualidade fisiológica de sementes de soja convencional e RR associada ao conteúdo de lignina. 134p. Thesis (Doutorado em Produção e Tecnologia de sementes) – Universidade Federal de Lavras, Lavras.

Gris, C. F.; Von Pinho, E.V. de R.; Andrade, T.; Baldoni, A.; Carvalho, M.L. de M. (Mar./Abr. 2010). Qualidade fisiológica e teor de lignina no tegumento de sementes de soja convencional e transgênica RR submetidas a diferentes épocas de colheita. *Ciência e Agrotecnologia*, Lavras, v. 34, n. 2, p. 374-381.

Hartwig, E.E.; Potts, H.C. (May/June 1987). Development and evaluation of impermeable seed coats for preserving soybean seed quality. *Crop Science*, Madison, v.27, n.3, p.506-508.

Hrazdina, G. & Jensen, R. A. (1992). Spatial organization of enzymes in plant metabolic pathways. *Annual Review of Plant Physiology and Plant Molecular Biology*, 43: 241-267.

International Service For The Acquisition Of Agri-Biotech Applications – ISAAA. (2009). *Brief 38-2009:* Global Status of Commercialized Biotech/GM Crops: 2009. Available at HTTP: http://www.isaaa.org/resources/publications/briefs/38/executivesummary/default.html

José, S.C.B.R.; Pinho, É.V.R. von; Pinho, R.G. von; Silveira, C.M. da. (Set./out. 2004). Tolerância de sementes de linhagens de milho a alta temperatura de secagem. *Ciência e Agrotecnologia*, Lavras, v.28, n.5, p.1107-1114.

Jung, H.G., Allen, M.S. (1995). Characteristics of plants cell walls affecting intake and digestibility of forages by ruminants. *Journal Animal Science*, Champaign, v.73, n.9, p.2774-2790.

Kruse, N.D.; Trezzi, M.M.; Vidal, R.R. (2000). Herbicidas inibidores da EPSPS: revisão de literatura. *Revista Brasileira de Herbicidas*, Brasília, v.1, n.2, p.139- 146.

Kuiper, H.A.; Kleter, G.A.; Noteborn, H.P.J.M.; Kok, E.J. (Dec. 2001). Assessment of the food safety issues related to genetically modified foods. *The Plant Journal*, Oxford, v.27, n.6, p.503-528.

Lacerda, A.L. de S.; Matallo, M.B. (2008). Verificação do ácido chiquímico em soja geneticamente modificada. In: Reunião Anual da SOCIEDADE BRASILEIRA PARA O PRPGRESSO DA CIÊNCIA, 60, Campinas. *Anais...* Campinas: UNICAMP, 2008. Available at: http://www.sbpcnet.org.br/livro/60ra/resumos/resumos/R2708-1.html>. Accessed on: Mar. 11, 2009.

Lewis, N.G., Yamamoto, E. (1990). Lignin: occurrence, biogenesis and biodegradation. *Ann. Rev. Plant Physiol. Plant Mol. Biol.*, Palo Alto, v.41, p.455-496.

Lin, S.S.; Severo, J.L. (1982). Efeito e atraso da colheita sobre a qualidade da semente e rendimento de soja (Glycine max (L.) Merrill). *Agronomia Sulriograndense*, Porto Alegre, v.18, n.1, p.37-46.

Machado, J.C.(2000). *Tratamento de sementes no controle de doenças*. Lavras: LAPS/UFLA/FAEPE. 138p.

Marcos Filho, J.(1999). Teste de envelhecimento acelerado. In: KRZYZANOWSKI, F.C.; VIEIRA, R.D.; FRANÇA NETO, J.B. (Ed.). *Vigor de sementes:* conceitos e testes. Londrina: ABRATES, 1999. cap.3, p.1-24.

Marcos Filho, J.; Cicero, S.M.; Silva, W.R. da. (1987). *Avaliação da qualidade da semente.* Piracicaba: FEALQ. 230p.

Martins, L.A.M. (1989). *Avaliação da área habilidade genética do caráter semente dura de linhagens melhoradas de soja (Glycine max* (L.) Merrill). 119f. Dissertation (Mestrado em Fitotecnia)-Universidade Federal de Viçosa, Viçosa, MG.

Mcdougall, G. J., Morrison, I. M., Stewart, D., Hillman J. R. (Feb. 1996). Plant cell walls dietary fibre: range, structure, processing and function. *Journal Science and Food Agriculture*, Londres, v. 70, n. 2, p. 133-150.

Mckinney, R.H. (1923). Influence of soil temperature and moisture on infection of wheat seedlings by *Helminthosporium sativum. Journal of Agricultural Research*, Washington, v.26, n.3, p.195-218.

Menezes, M. de. (2008). *Aspectos genéticos associados à qualidade fisiológica de sementes de soja.* 112p. Thesis (Doutorado em Fitotecnia)-Universidade Federal de Lavras, Lavras.

Menezes, M.; Von Pinho, E.V. de R.; Roveri José, S.C.B.; Baldoni, A.; Mendes, F.F. (December 2009). Aspectos químicos e estruturais da qualidade fisiológica de sementes de soja. *Pesquisa Agropecuária Brasileira*, Brasília, v. 44, n. 12, p. 1716-1723.

Nodari, R.O.; Destro, D. (2006). *Relatório sobre a situação de lavouras de soja da região de Palmeira das Missões, RS, safra 2001/2002, cultivadas com cultivares convencionais e com cultivares transgênicas:* notícias no AgirAzul. Available at: <http://www.agirazul.com.br/123/noticias/000000a3.htm>. Accessed on: Apr. 22, 2006.

Panobianco, M. (1997). *Variação na condutividade elétrica de sementes de diferentes genótipos de soja e relação com o conteúdo de lignina no tegumento.* 59p. Dissertation (Mestrado em Agronomia)-Universidade Estadual Paulista, Jaboticabal.

Panobianco, M.; Vieira, R. D.; Krzyzanowski, F. C.; França Neto, J.B. (1999). Electrical conductivity of soybean seed and correlation with seed coat lignin content. *Seed Science Technology*, Zurich, v.27. n.3, p.945-949.

Peske, S.T.; Pereira, L.A.G. (Jun. 1983). Tegumento da semente de soja. *Tecnologia de Sementes*, Pelotas, v.6, n.1/2, p.23-34.

Resende, M. L. V.; Salgado, S. M.; Chaves, Z. M. (Mar./abr. 2003). Espécies ativas de oxigênio na resposta de defesa de plantas a patógenos. *Fitopatologia Brasileira*, Brasília, v. 28, n. 2, p. 123-130.

Rocha, V.S. (1982). *Avaliação da qualidade fisiológica de sementes de genótipos de soja (Glycine max (L.) Merrill), em três épocas de colheita.* 109f. Dissertation (Mestrado em Fitotecnia)-Universidade Federal de Viçosa, Viçosa, MG.

Salisbury, F. K. & Ross, C. W. (1992). *Plant Physiology.* Wadsworth, Belmont.

Salisbury, F. K. & Ross, C. W. (1992). *Plant Physiology.* Wadsworth, Belmont.

Sanino, F.; Filazzola, M. T.; Violante, A. (1999). Fate of herbicides influenced by biotic and abiotic interactions. *Chemosphere*, [S.l.], v. 39, n. 2, p. 333-341.

Silva, D.J. (1981). *Análise de alimentos:* métodos químicos e biológicos. Viçosa, MG: UFV. 166p.

Silva, M. A. D. (2003). *Morfologia da testa e potencial fisiológico de sementes de soja.* 84p. Thesis (Doutorado em Produção e Tecnologia de sementes) – Universidade Federal de Lavras, Lavras.

Simões, C.M. O.; Spitzer, V. (2004). *Óleos voláteis.* In: Simões, C.M. O. et al. (Org.). Farmacognosia da planta ao medicamento. 5. ed. Porto Alegre: UFRGS; Florianópolis: UFSC, cap. 18, p. 467-496.

Strautman, B. (2007). Manganese affected by glyphosate. *Western Producer.* http://www.gefreebc.org/gefree_tmpl.php?content=manganese_glyphosate

Swanson, B. G.; Hughes, J. S.; Rasmussen, H. (1985). Seed microstructure: review of water imbibition in legumes. *Food Microstructure,* Chicago, v. 4, p. 115-124.

Taiz, L. & Zeiger, E. (1998). *Plant Physiology.* Sinauer Associates, Sunderland. 792p.

Tavares, D.Q.; Miranda, M.A.C.; Umino, C.Y.; Dias, G.M. (Jan./mar. 1987). Características estruturais do tegumento de sementes permeáveis e impermeáveis de linhagens de soja, *Glycine max* (L.) Merrill. *Revista Brasileira de Botânica,* São Paulo, v.10, n.1, p.147-153.

Vieira, R.D.; Krzyzanowski, F.C. (1999). *Teste de condutividade elétrica.* In: KRZYANOWSKI, F.C.; VIEIRA, R.D.; FRANÇA NETO, J.B. (Ed.). Vigor de sementes: conceitos e testes. Londrina: ABRATES. cap.4, p.1-26.

Woodstock, L.W. (Feb. 1988). Seed imbibition: a critical period for successful germination. *Journal Seed Technology,* Springfield, v.12, n.1, p.1-15.

Xu, Z. et al. (Oct. 2009). Comparative genome analysis of lignin biosynthesis gene families across the plant kingdom. *BMC Bioinformatics,* New York, v. 10, n. 11, p. 1-15. Supplement.

Zobiole, L.H.S., Oliveira, R.S., Visentainer, J.V., Kremer, R.J., Bellaloui, N., Yamada, T. (2010). Glyphosate affects seed composition in glyphosate-resistant soybean. *J. Agric. Food Chem.* 58, 4517–4522.

2

A Comparative Study of the Chelating Effect Between Textured Soya Aqueous Extract and EDTA on Fe^{3+}, Pb^{2+}, Hg^{2+}, Cd^{2+} and Ni^{2+} Ions

Guajardo Jesús, Morales Elpidio, López Francisco,
Quintero Cristina, Compean Martha, Noriega María-Eugenia,
González Jesús and Ruiz Facundo
Universidad Autónoma de San Luis Potosí
Mexico

1. Introduction

Metal pollution of soils, water, foods, and the environment is a grave problem. Various in-situ and ex-situ remediation techniques have been employed, e.g., solidification, stabilization, flotation, soil ashing, electroremediation, bioleaching, and phytoremediation (Mulligan, 2001). One remediation technique is ex-situ soil washing using chelating agents. The soil is removed from the site, treated in a closed reactor with the chelating agent, and returned to the site after separation of the extraction solution that now contains the extracted heavy metals (Peters & Hazard,1999). The problem is that the used chelating agent is not a natural compound. For that reason we propose the use of the textured soya extract, which is environmentally friendly, as a natural chelating agent.

EDTA (ethylenediamine tetraacetic acid) and its salts are substituted diamines. HEDTA (hydroxyethyl ethylenediamine triacetic acid) and its trisodium salt are substituted amines. These ingredients function as chelating agents in cosmetic formulations. The typical concentration of use of EDTA is less than 2%, with the other salts in current use at even lower concentrations. The lowest dose reported to cause a toxic effect in animals was 750 mg/kg/day.

These chelating agents are cytotoxic and weakly genotoxic, but not carcinogenic. Oral exposures to EDTA produced adverse reproductive and developmental effects in animals. Clinical tests reported no absorption of an EDTA salt through the skin. These ingredients are likely, however, to affect the passage of other chemicals into the skin because they will chelate calcium. Exposure to EDTA in most cosmetic formulations, therefore, would produce systemic exposure levels well below those seen to be toxic in oral dosing studies. Exposure to EDTA in cosmetic formulations that may be inhaled, however, was a concern. An exposure assessment done using conservative assumptions predicted that the maximum EDTA dose via inhalation of an aerosolized cosmetic formulation is below that shown to produce reproductive or developmental toxicity. Because of the potential to increase the penetration of other chemicals, formulators should continue to be aware of this when combining these ingredients with ingredients that previously have been determined to be safe, primarily because they were not significantly absorbed. Based on the available data, the Cosmetic Ingredient Review Expert Panel found that these ingredients are safe as used in cosmetic formulations.

Ethylenediaminetetraacetic acid (EDTA) is a very effective chelating agent but has the disadvantage that is quite persistent in the environment owing to its low biodegradability. For that reason different chelating agents were investigated, such as [S,S,]-ethylenediaminedisuccinic acid, iminodisuccinic acid, methylglycine diacetic acid, etc. but the problem is the dependence of the pH on the extraction efficiency. (Tandy et al., 2004)

Major industrial processes involve the sequestration of metal ions in an aqueous solution. In the textile industry, this prevents metal ion impurities from modifying colors of dyed products. In the pulp and paper industry, EDTA inhibits the ability of metal ions, especially Mn^{2+}, to catalyze disproportionate amounts of hydrogen peroxide, which is used in "chlorine-free bleaching." Similarly, EDTA is added to some foods as a preservative or stabilizer to prevent a catalytic oxidative discoloration which is catalyzed by metal ions.

Oral exposures have been noted to cause reproductive and developmental effects (Elliot & Brown, 1989). The same study by Lanigan also found that both dermal exposure to EDTA in most cosmetic formulations and inhalation exposure to EDTA in aerosolized cosmetic formulations would produce systemic effects below those seen to be toxic in oral dosing studies (Lanigan & Yamarik, 2002).

A crucial factor to be considered in comparing studies on chelating agent is the pH of the extraction solution. While extraction was investigated at various pH values in some studies (Elliot & Brown, 1989 ,; Pichtel, 1998; Pichtel, 1997; Kim, 2003; Ghestem, 1998), some only stated the pH of the solution (Reed, 1996; Cline, 1995; Van Benschoten, 1997), while others did not consider pH at all (Pichtel, 2001;). In general, the lower the pH of the chelating agent solution, the greater is the extraction efficiency of the toxic metals.

The history and chemistry of the industrial use of natural products and their derivatives have a rich technological tradition. Many modern products, such as plastics, fuels, chemical intermediates and fibers, find their origins in natural products derived from plants and animals. Given the recent social emphasis on the environment and resource renewability, utilizing natural materials as potential resources for industrial products receives a ready welcome. Among the most versatile of raw materials is the soybean. (Liu, 1997)

Together, the oil and protein contents of dry soybeans account for about 60% of the weight; protein being 40% and oil 20%. The remainder consists of 35% carbohydrate and about 5% ash. Most soy protein is a relatively heat-stable storage protein. This heat stability enables the manufacture of soy food products requiring high temperature cooking, such as tofu, soy milk and textured vegetable protein (soy flour).

This article focuses on the application of natural "green" textured soya extract as a substitute for EDTA in its role as a metals-sequestering agent in foods.

2. Antecedents

2.1 What is a chelating agent?

The word chelation is derived from Greek, meaning "claw." The ligands lie around the central atom like the claws of a lobster.

The IUPAC definition of chelation is the formation or presence of two or more separate bindings between a polydentate (multiple bonded) ligand and a single central atom. Usually these ligands are organic compounds and are called chelants, chelators, chelating agents, or sequestering agents. (IUPAC)

A Comparative Study of the Chelating Effect Between Textured Soya Aqueous Extract and EDTA on Fe³⁺, Pb²⁺,
Hg²⁺, Cd²⁺ and Ni²⁺ Ions

27

The ligand forms a chelate complex with the substrate. Chelate complexes are contrasted with coordination complexes composed of monodentate ligands, which form only one bond with the central atom. (Morgan & Drew, 1920)

The terms bidentate (or didentate), tridentate, tetradentate,... multidentate are used to indicate the number of potential binding sites of the ligand, at least two of which must be used by the ligand in forming a "chelate". For example, the bidentate ethylenediamine forms a chelate with **Cu (I)** in which both nitrogen atoms of ethylenediamine are bonded to copper. (The use of the term is often restricted to metallic central atoms). (Kramer, Cotter-Howells, Charnock, Baker & Smith. 1996)

Chelants, according to ASTM-A-380, are "chemicals that form soluble, complex molecules with certain metal ions, inactivating the ions so that they cannot normally react with other elements or ions to produce precipitates or scale".

2.2 The chelate effect

The increased stability of complexes containing chelating ligands over those containing comparable monodentate ligands can be envisaged as having the following physical basis. Suppose we have a metal ion in solution, and we attach to it a monodentate ligand, followed by a second monodentate ligand, figure 1. These two processes are completely independent of each other. But suppose we have a metal ion and we attach to it one end of a chelating ligand (it is reasonable to assume that when we put a chelate ligand onto a metal, it happens in a stepwise fashion, i.e. one end attaches first and then the other end). The point is that the attachment of the second end of the chelate is now no longer an independent process: once one end is attached, the other end, rather than floating around freely in solution, is anchored by the linking group in reasonably close proximity to the metal ion, and is therefore more likely to join onto it than a comparable monodentate ligand would be.

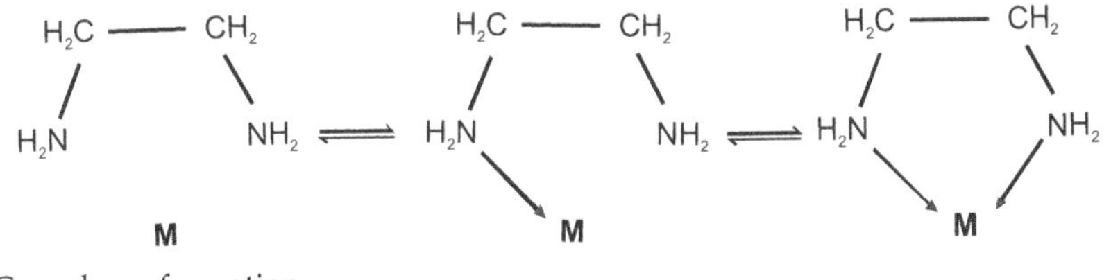

Fig. 1. Complexes formation.

The figure 2 shows the EDTA ligand binding to a central copper ion.

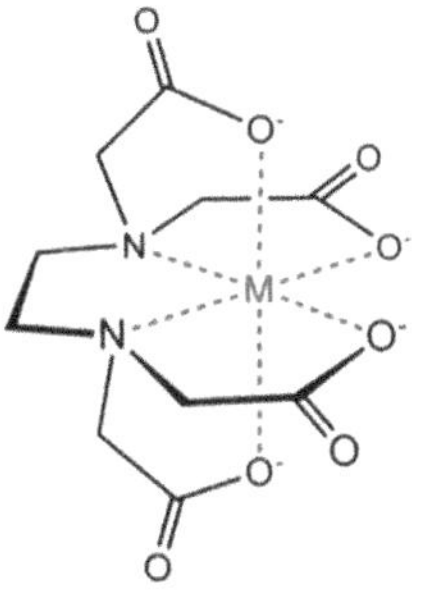

Fig. 2. Copper ion complexes with EDTA.

Amino acids are classified into different ways base don polarity, structure, nutricional requirement, metabolic fate, etc.

Generally used classification is based on polarity. Based on polarity amino acids are classified into four groups.

- *Non-polar amino acids.- They have equal number of amino and carboxyl groups and are neutral. These amino acids are hydrophobic and have no charge on the 'R' group. The amino acids in this group are alanine, valine, leucine, isoleucine, phenyl alanine, glycine, tryptophan, methionine and proline.*

- *Polar amino acids with no charge.- These amino acids do not have any charge on the 'R' group. These amino acids participate in hydrogen bonding of protein structure. The amino acids in this group are - serine, threonine, tyrosine, cysteine, glutamine and aspargine.*

- *Polar amino acids with positive charge.- Polar amino acids with positive charge have more amino groups as compared to carboxyl groups making it basic. The amino acids, which have positive charge on the 'R' group, are placed in this category. They are lysine, arginine and histidine.*

- *Polar amino acids with negative charge.- Polar amino acids with negative charge have more carboxyl groups than amino groups making them acidic. The amino acids, which have negative charge on the 'R' group are placed in this category. They are called as dicarboxylic mono-amino acids. They are aspartic acid and glutamic acid.*

Chelates of glycine with cations such as iron, zinc and copper have been fully studied. The chelates usually contain two moles of ligand (glycine) and one mol of metal as demonstrated in the figure 3.

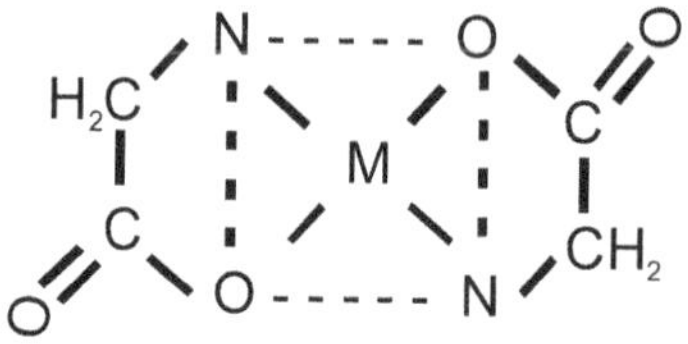

Fig. 3. Chelate of glycine with some metal M.

Consider the two equilibriums, in an aqueous solution, between the copper (II) ion, Cu^{2+} and ethylenediamine (en) on the one hand and methylamine, $MeNH_2$ on the other.

$$Cu^{2+} + en \rightleftharpoons [Cu(en)]^{2+} \tag{1}$$

$$Cu^{2+} + 2\,MeNH_2 \rightleftharpoons [Cu(MeNH_2)_2]^{2+} \tag{2}$$

In (1) the bidenate ligand ethylene diamine forms a chelate complex with the copper ion. Chelation results in the formation of a five-member ring. In (2) the bidentate ligand is replaced by two monodentate methylamine ligands of approximately the same donor power, meaning that the enthalpy of formation of $Cu-N$ bonds is approximately the same in the two reactions. Under conditions of equal copper concentrations and when the concentration of methylamine is twice the concentration of ethylenediamine, the concentration of the complex (1) will be greater than the concentration of the complex (2). The effect increases with the number of chelate rings so the concentration of the EDTA complex, which has six chelate rings, is much higher than a corresponding complex with two monodentate nitrogen donor ligands and four monodentate carboxylate ligands. Thus, the phenomena of the chelate effect are a firmly established empirical fact.

A Comparative Study of the Chelating Effect Between Textured Soya Aqueous Extract and EDTA on Fe^{3+}, Pb^{2+}, Hg^{2+}, Cd^{2+} and Ni^{2+} Ions

29

The thermodynamic approach to explaining the chelate effect considers the equilibrium constant for the reaction: the larger the equilibrium constant, the higher the concentration of the complex.

The formation of a chelant compound is an equilibrium reaction as shown in the reaction (3)

$$a M^{n+} + b L \iff c ML \tag{3}$$

$$\text{Metal} \quad \text{Ligand} \quad \text{Metal-chelate}$$

The reaction rates of the forward and reverse reactions are generally not zero but, being equal; there are no net changes in any of the reactant or product concentrations. Since forward and backward rates are equal:

$$k_1 [M^{n+}]^a [L]^b = k_2 [ML]^c \tag{4}$$

and the ratio of the rate constants is also a constant, now known as an equilibrium constant.

$$K = \frac{[ML]^c}{[M^{n+}]^a [L]^b} \tag{5}$$

The concentration of ligand does not change during the reaction. For that reason the equilibrium constant can be expressed only in function of metal ion and metal-complex, as showing in the equation 6.

$$K = \frac{[ML]^c}{[M^{n+}]^a} \tag{6}$$

2.3 Common chelating agents

There are many chelating agents used in the industry as Na, Ca-ethylenediaminetetraacetic (EDTA), diethylenetriaminepentaacetic acid (DTPA), nitriloacetic acid, ethylene glycol-bis8aminoethyl)tetraacetic acid (EGTA), D,L-mercaptosuccinic acid (MSA), meso-2-3-dimercaptopropanesuccinic acid (DMSA), D,L-2,3-dimercaptopropane-1-sulfonic acid (DMPS), penicillamine (PA), N-acetylpenicillamine (NAPA), vitamins as: thiamine (B1), pyridoxine (B6), cobalim (B12) and ascorbic acid, and many more. The most common is EDTA.

2.4 Naturals chelating agents

Virtually all biochemicals exhibit the ability to dissolve certain metals cations. Thus, proteins, polysaccharides, and polynucleic acids are excellent polydentate ligands for many metal ions. In addition to these adventitious chelators, several biomolecules are produced to specifically bind certain metals. Histidine, malate and phytochelatin are typical chelators used by plants. (U Kramer, 1996; Jurandir, 2006 & Suk-Bomg Há, 1999)

Virtually all metalloenzymes feature metals that are chelated, usually to peptides or cofactors and prosthetic groups (Lippard & Berg, 1994). Such chelating agents include the porphyrin in hemoglobin and chlorophyll. Many microbial species produce water-soluble pigments that serve as chelating agents, termed sideropho. For example, species of *Pseudomonas* are known to secrete pycocyanin and pyoverdin that bind iron. Enterobactin, produced by E. coli, is the strongest chelating agent known.

In earth science, chemical weathering is attributed to organic chelating agents, *e.g.* peptides and sugars that extract metal ions from minerals and rocks. (Michael) Most metal complexes in the environment and in nature are bound in some form of chelate ring, *e.g.* with a humic acid or a protein. Thus, metal chelates are relevant to the mobilization of metals in the soil, the uptake and the accumulation of metals into plants and micro-organisms. Selective chelating of heavy metals is relevant to bioremediation *e.g.* removal of ^{137}Cs from radioactive waste. (Prasad, 2001)

2.5 Applications

Chelators are used in chemical analysis as water softeners, and are ingredients in many commercial products such as shampoos and food preservatives. Citric acid is used to soften water in soaps and laundry detergents. A common synthetic chelator is EDTA. Phosphona are also well known chelating agents. Chelators are used in water treatment programs and specifically in steam engineering, e.g., boiler water treatment system.

Chelation therapy is the use of chelating agents to detoxify poisonous metal agents such as mercury, arsenic, and lead by converting them to a chemically inert form that can be excreted without further interaction with the body, and was approved by the U.S. Food and Drug Administration in 1991. In alternative medicine, chelation is used as a treatment for autism, though this practice is controversial due to an absence of scientific plausibility, lack of FDA approval, and its potentially deadly side-effects. (Doja & Can, 2006).

Though they can be beneficial in cases of heavy metal poisoning, chelating agents can also be dangerous. The U.S. CDC reports that use of disodium EDTA instead of calcium EDTA has resulted in fatalities due to hypocalcemia. (U. S. Center for Disease Control)

Homogeneous catalysts are often chelated complexes. A typical example is the ruthenium (II) chloride chelated with BINAP (a bidentate phosphine) used in e.g. Noyori asymmetric hydrogenation and asymmetric isomerization. The latter has the practical use of manufacture of synthetic mentol.

Products such as Evapo-Rust are chelating agents sold for the removal of rust from iron and steel.

2.6 Chemical composition of the soybean seed

Together, oil and protein content account for about 60% of dry soybeans by weight; protein at 40% and oil at 20%. The remainder consists of 35% carbohydrate and about 5% ash. Soybean cultivars comprise approximately 8% seed coat or hull, 90% cotyledons and 2% hypocotyl axis or germ.

Most soy protein is a relatively heat-stable storage protein. This heat stability enables soy food products requiring high temperature cooking, such as tofu, soy milk and textured vegetable protein (soy flour) to be made.

The principal soluble carbohydrates of mature soybeans are the disaccharide sucrose (range 2.5–8.2%), the trisaccharide raffinose (0.1–1.0%) composed of one sucrose molecule connected to one molecule of galactose, and the tetrasaccharide stachyose (1.4 to 4.1%) composed of one sucrose connected to two molecules of galactose. While the oligosaccharides raffinose and stachyose protect the viability of the soy bean seed from desiccation (see above section on physical characteristics) they are not digestible sugars and therefore contribute to flatulence and abdominal discomfort in humans and other monogastric animals; compare to the disaccharide trehalose.

A Comparative Study of the Chelating Effect Between Textured Soya Aqueous Extract and EDTA on Fe^{3+}, Pb^{2+}, Hg^{2+}, Cd^{2+} and Ni^{2+} Ions

31

Since soluble soy carbohydrates are found in the whey and are broken down during fermentation, soy concentrate, soy protein isolates, tofu, soy sauce, and sprouted soy beans are without flatus activity. On the other hand, there may be some beneficial effects to ingesting oligosaccharides such as raffinose and stachyose, namely, encouraging indigenous bifidobacteria in the colon against putrefactive bacteria.

The insoluble carbohydrates in soybeans consist of the complex polysaccharides cellulose, hemicellulose and pectin. The majority of soybean carbohydrates can be classed as belonging to dietary fiber.

The following Table 1 shows the composition of mature, raw soybean seeds.

Energy	1,866 kJ (446 kcal)
Carbohydrates	30.16 g
Sugars	7.33 g
Dietary fiber	9.3 g
Fat	19.94 g
Saturated	2.884 g
monounsaturated	4.404 g
polyunsaturated	11.255 g
Protein	36.49 g
Tryptophan	0.591 g
Threonine	1.766 g
Isoleucine	1.971 g
Leucie	3.309 g
Lysine	2.706 g
Methionine	0.547 g
Phenylalanine	2.122 g
Tyrosine	1.539 g
Valine	2.029 g
Arginine	3.153 g
Histidine	1.097 g
Alanine	1.915 g
Aspartic acid	5.112 g
Glutamic acid	7.874 g
Glycine	1.880 g
Proline	2.379 g
Serine	2.357 g
Water	8.54 g
Vitamin A equiv	1 µg
Vitamin B_6	0.377 mg
Vitamin B_{12}	0 µg
Vitamin C	6.0 mg
Vitamin K	47 µg
Calcium	277 mg
Iron	15.70 mg
Magnesium	280 mg
Phosphorus	704 mg
Potassium	1797 mg
Sodium	2 mg
Zinc	4.89 mg

Source: USDA Nutrient database.

Table 1. Composition of soybean, mature, rawNutritional value per 100 g (3.5 oz)

2.7 How soybeans are used

When the farmer sells soybeans to a grain dealer, the beans may then go to a number of ultimate destinations. When processed, a 60-pound bushel will yield about 11 pounds of crude soybean oil and 47 pounds of soybean meal. Soybeans are about 18% oil and 38% protein. Because soybeans are high in protein, they are a major ingredient in livestock feed. Most soybeans are processed for their oil and protein for the animal feed industry. A smaller percentage is processed for human consumption and made into products including soy milk, soy flour, soy protein, tofu and many retail food products. Soybeans are also used in many non-food (industrial) products.

Fuel for diesel engines can be produced from soybean oil with simple processing. Soy biodiesel is cleaner burning than petroleum-based diesel oil. Its use reduces particulate emissions, and it is non-toxic, renewable and environmentally friendly. Soy crayons made by the Dixon Ticonderoga Company replace the petroleum used in regular crayons with soy oil making them non-toxic and safer for children. Candles made with soybean oil burn longer but with less smoke and soot.

Soy oil produces an environmentally friendly solvent that safely and rapidly removes oil from creeks, streams and shorelines without harming people, animals, and the environment. Soy is an ingredient in many industrial lubricants, solvents, cleaners, and paints. Soy ink is superior to petroleum-based inks because soy ink is not toxic, is renewable and also environmentally friendly. Furthermore, it cleans up easily. Soy-based lubricants are as good as petroleum-based lubricants, but can withstand higher heat. More importantly, they are non-toxic, renewable and environmentally friendly. Soy-based hydraulic fluid and rail flange lubricants are among the more recent products developed with check-off funds.

Soy-based foams are currently being developed for use in coolers, refrigerators, automotive interiors and even footwear. Beginning in October 2007, Ford Mustangs rolled off the production line with soy flexible foam in the seats. (2009 Annual Report of the North Carolina Soybean Producers Association)

2.8 Textured soya

Textured or textures vegetable protein (TVP), also known as textured soya protein (TSP), soy meat, or soya meat is a meat analogue or nutritious meat extender made from defatted soy flour, a by-product of extracting soybean oil. It is quick to cook, with a protein content equal to that of the meat, and contains no fat. (Riaz, 2006)

TVP is made from a mixture of proteins extracted primarily from soybeans, but also cotton seeds, wheat, and oats. It is extruded into various shapes (chunks, flakes, nuggets, grains, and strips) and sizes, exiting the nozzle while still hot and expanding as it does so. (Foote, 1996)

TVP can be made from soy flour or concentrate, containing 50% and 70% soy protein respectively, and is relatively flavorless. Both require rehydration before use, sometimes with flavoring added in the same step. TVP is extruded, causing a change in the structure of the soy protein which results in a fibrous spongy matrix that is similar in texture to meat. In its dehydrated form TVP has a shelf life of longer than a year, but will spoil within several days after being hydrated. In its flaked form TVP can be used similarly to ground meat. (Hoogenkamp & Wallingford, Oxon, 2005; Endres, 2001)

A Comparative Study of the Chelating Effect Between Textured Soya Aqueous Extract and EDTA on Fe^{3+}, Pb^{2+}, Hg^{2+}, Cd^{2+} and Ni^{2+} Ions

33

3. Materials and synthesis

3.1 Reagents

$Fe(NO_3)_3$, $NiCl_2$, $CuSO_4$, $HgCl_2$, $CdSO_4$, $Pb(NO_3)_2$ and EDTA (Etilenediaminetetraacetic acid), purchased from Sigma-Aldrich, were used without any further purification. Textured soya was purchased. Mili-Q water (18.2 Ω) was used throughout the experiment.

3.2 Characterization

The amount of metallic ions present in the solutions was determined by using a conductivity meter.

3.3 Determination of the chelating agent in textured soya extract

The Biuret test is a chemical test used for detecting the presence of peptide bonds. In the presence of peptides, a copper (II) ion forms a violet-colored complex in an alkaline solution. Several variants on the test have been develop

In order to find the chelating component in the textured soya extract, first a textured soya extract was prepared by heating to boiling point 1000 ml of deionizer water with 30 grams of textured soy for 20 minutes. 10 ml of textured soya extract is treated with an equal volume of 1% strong base (sodium or potassium hydroxide most often) followed by a few drops of aqueous copper (II) sulfate. The solution turns violet, for that reason we can affirm that proteins are present in the textured soya extract and these are the chelating agents.

A Fehling test was made too, and the Fehling reaction was positive, in which the green color characteristic of mono-saccharides was obtained. For that reason we can affirm that mono-saccharides present in the textured soya extract are present but the amount is not significant (< 3%).

3.4 Experimental method

First, it was necessary to find a concentration of textured soya equivalent to an EDTA solution $5x10^{-4}$ (the maximum concentration permitted in foods). For that reason we prepared six solutions of $CuSO_4$ with concentrations 0.05M, 0.1M, 0.15M, 0.2M, 0.25M and 0.3M. The conductivity of each one was then measured. Next, we mixed 1 ml of each $CuSO_4$ solution with 10 ml of an EDTA solution $5x10^{-4}$ M and we measured the conductivity of each one. Several textured soya aqueous solutions were prepared by dissolving 1 grams, 2 grams, 3 grams, and 5 grams of textured soya, each one in 100 ml of water, and heating them to boiling point for 10 minutes. The fiber was then separated by filtration. Afterwards, we mixed 1 ml of each solution of $CuSO_4$ with 10 ml of each prepared textured soya extract solution and measured the conductivity of each sample.

With the aim of studying the comparative chelating effect between textured soya extract and EDTA on some metals, we prepared five different aqueous solutions of each metal ion, Fe^{3+}, Pb^{2+}, Hg^{2+}, Cd^{2+}, Ni^{2+} and Cu^{2+} with different concentrations, with 0.01, 0.03, 0.05, 0.07 and 0.1 grams of each salt dissolved in 10 ml of deionizer water. Then we measured their conductivity. An EDTA aqueous solution of $5x10^{-4}$M was prepared. A solution of 15 grams of textured soya in 500 ml of deionizer water was heated to boiling point for 10 minutes. Afterwards, we measured the conductivity and ppm (parts of million) of each ion solution, chelant solution of textured soya extract and EDTA. In order to determine the chelating capacity of EDTA and textured soya extract, we added 0.01, 0.03, 0.05, 0.07 and 0.1 grams of each salt in 10 ml of EDTA solution, and then in the same form in 10 ml of textured soya extract solution and we measured the conductivity of each one, using a conductivity meter. All measurements were made at room temperature and at average room pressure, and pH 7.

4. Results and discussion

Table 2 shows the conductivity and parts per million of EDTA solution (5×10^{-4} M), measured with the conductivity meter, and the four different solutions of textured soya extract prepared.

	1 gram textured soya	2 grams textured soya	3 grams textured soya	5 grams textured soya	EDTA (5×10^{-4} M)
Conductivity (µs)	459.1	454.7	836.6	1191	63.6
ppm	308.5	303	564.6	820.3	40.51

Table 2. Conductivity y ppm of EDTA and textured soya extract solutions

Table 3 shows the resulting conductivity after mixing each of the $CuSO_4$ solutions with the EDTA solution and the four textured soy extract solutions. These results are the differences between the measurement of the mixture of the $CuSO_4$ solution with the soy chelating solution and the pure extraction solution.

Concentration $CuSO_4$ (M)	1 gram textured soya (µs)	2 grams textured soya (µs)	3 grams textured soya (µs)	5 grams textured soya (µs)	EDTA (µs)
0.025	481.8	402	401.4	443	820
0.05	958.9	645.3	853.4	835	1350.4
0.1	2157.9	1857.3	1619.4	1685	2172.4
0.15	2569.9	2969.3	2479.4	2614	3057.4
0.2	3195.9	3503.3	2876.4	3340	3682.4
0.25	3407.9	4310.3	3968.4	3952	4200.4
0.3	3780.9	4836.3	4590.4	4631	4919.4

Table 3. Conductivity of EDTA solution and textured soya extract solutions with the prepared $CuSO_4$ solutions

Figure 4 shows graphically the results of the Table 3.
From these results, we can conclude that is necessary to prepare the textured soya extract solution by using 2 or 3 grams of textured soy in 100 ml of water, heating it to boiling point for 10 minutes and separating out the fiber by filtration.
Table 4 contains the conductivity and ppm of the aqueous EDTA solution and the aqueous textured soya extract using 3 grams of textured soya in 100 ml of deionizer water we prepared.

	Textured soya extract	EDTA (5×10^{-4}M)
Conductivity (µs)	1414	150.3
ppm	981	67.06

Table 4. conductivity and ppm of EDTA and textured soy extract solutions.

A Comparative Study of the Chelating Effect Between Textured Soya Aqueous Extract and EDTA on Fe^{3+}, Pb^{2+}, Hg^{2+}, Cd^{2+} and Ni^{2+} Ions

35

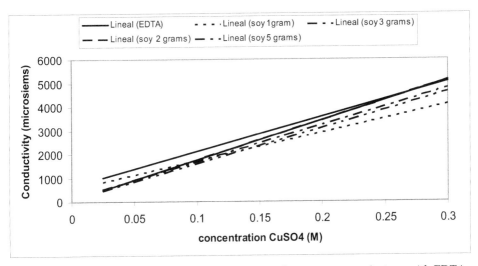

Fig. 4. Conductivity of the mixture of different textured soya extracts solutions with EDTA solution with the $CuSO_4$ solutions .

Table 5 shows the results of the conductivity and ppm of five aqueous Pb^{2+} solution prepared dissolved 0.01, 0.03, 0.05, 0.07 and 0.1 grams of $Pb(NO_3)_2$ in 10 ml of deionizer water, each one. In a similar process were prepared five aqueous solutions EDTA-Fe^{3+} dissolved 0.01, 0.03, 0.05, 0.07 and 0.1 grams of $Pb(NO_3)_2$ in 10 ml of aqueous EDTA solution and in the same way were prepared five aqueous solutions of textured soya extract-Pb^{2+} dissolved 0.01, 0.03, 0.05, 0.07 and 0.1 grams of $Pb(NO_3)_2$ in 10 ml of textured soya extract.

Grams $Pb(NO_3)_2$	Pb^{2+} aqueous solutions ppm (µs)		Textured soya extract-Pb^{2+} Ppm (µs)		EDTA- Pb^{2+} $(5x10^{-4}M)$ ppm (µs)	
0.01	472.45	704.54	116	160	629	907.7
0.03	1193.35	1703.34	572	762	933	1328.7
0.05	2020.35	2764.34	1332	1720	2363	2323.7
0.07	2759.35	3687.34	2016	2569	2519	3354.7
0.1	4031.35	5203.34	3961	3815	3896	5013.7

Table 5. Conductivity and ppm of aqueous solutions: Pb^{2+} , EDTA- Pb^{2+} and textured soy extract-Pb^{2+}.

The conductivity and ppm of the mixture of EDTA and textured soya extract with aqueous Pb^{2+} solution, shown in the Table 5, are a result of subtracting the conductivity or ppm of the mixtures and conductivity and ppm from the EDTA and texture soya extract with $Pb(NO_3)_2$.

Figure 5 is a graphic representation of Table 5 results.

Table 6 shows the results of the conductivity and ppm of five aqueous Fe^{3+} solution prepared dissolved 0.01, 0.03, 0.05, 0.07 and 0.1 grams of $Fe(NO_3)_3$ in 10 ml of deionizer water, each one. In a similar process were prepared five aqueous solutions EDTA-Fe^{3+} dissolved 0.01, 0.03, 0.05, 0.07 and 0.1 grams of $Fe(NO_3)_3$ in 10 ml of aqueous EDTA solution and in the same way were prepared five aqueous solutions of textured soya extract-Fe^{3+} dissolved 0.01, 0.03, 0.05, 0.07 and 0.1 grams of $Fe(NO_3)_3$ in 10 ml of textured soya extract.

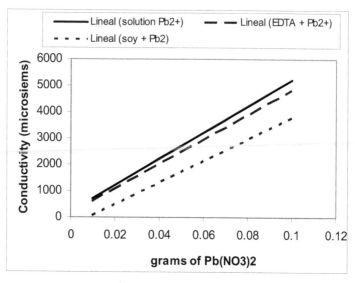

Fig. 5. Conductivity of aqueous solutions: Pb^{2+}, EDTA- Pb^{2+} and textured soya extract-Pb^{2+} .

Grams $Fe(NO_3)_3$	Fe^{3+} aqueous solutions ppm (μs)		Textured soya extract-Fe^{3+} Ppm (μs)		EDTA- Fe^{3+} (5×10^{-4}M) ppm (μs)	
0.01	932.64	1351.34	138	177	902.1	1286.7
0.03	2335.34	3159.34	995	1261	2097	2826.7
0.05	3910.34	5072.34	2785	3421	3573	4620.7
0.07	5078.34	6446.34	4431	5336	5008	6302
0.1	7329.34	8964.34	6065	7141	6900	8442

Table 6. Conductivity and ppm of aqueous solutions: Fe^{3+} , EDTA- Fe^{3+} and textured soya extract-Fe^{3+}.

The conductivity and ppm of the mixture of EDTA and textured soya extract with aqueous Fe^{3+} solution, shown in the Table 6, are a result of subtracting the conductivity or ppm of the mixtures and conductivity and ppm from the EDTA and texture soya extract with $Fe(NO_3)_3$. Figure 6 is a graphic representation of the Table 6 results.

Table 7 shows the results of the conductivity and ppm of five aqueous Cd^{2+} solution prepared dissolved 0.01, 0.03, 0.05, 0.07 and 0.1 grams of $CdSO_4$ in 10 ml of deionizer water, each one. In a similar process were prepared five aqueous solutions EDTA-Cd^{2+} dissolved 0.01, 0.03, 0.05, 0.07 and 0.1 grams of $CdSO_4$ in 10 ml of aqueous EDTA solution and in the same way were prepared five aqueous solutions of textured soya extract-Cd^{2+} dissolved 0.01, 0.03, 0.05, 0.07 and 0.1 grams of $CdSO_4$ in 10 ml of textured soya extract.

A Comparative Study of the Chelating Effect Between Textured Soya Aqueous Extract and EDTA on Fe³⁺, Pb²⁺, Hg²⁺, Cd²⁺ and Ni²⁺ Ions

37

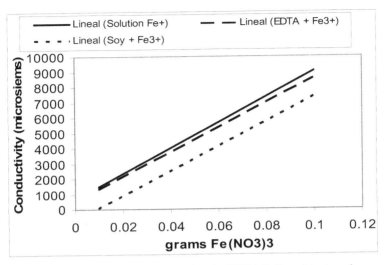

Fig. 6. Conductivity of the aqueous solution of: Fe^{3+}, $EDTA$-Fe^{3+} and textured soya extract-Fe^{3+}.

Grams CdSO₄	Cd²⁺ aqueous solutions ppm (µs)		Textured soya extract-Cd²⁺ Ppm (µs)		EDTA-Cd²⁺ (5x10⁻⁴M) Ppm (µs)	
0.01	393.75	591.54	82	110	439	643.4
0.03	948.55	1373.34	528	704	521.4	757.2
0.05	1346.35	1906.34	861	1135	1458	1947.7
0.07	1760.35	2442.34	1326	1301	2078	2807.7
0.1	2189.35	2979.34	1760	2256	2704	3584.7

Table 7. Conductivity and ppm of aqueous solutions: Cd²⁺, EDTA- Cd²⁺ and textured soya extract-Cd²⁺.

The conductivity and ppm of the mixture of EDTA and textured soya extract with aqueous Cd²⁺ solution, shown in the Table 7, are a result of subtracting the conductivity or ppm of the mixtures and conductivity and ppm from the EDTA and textured soya extract solutions with CdSO₄.

Figure 7 is a graphic representation of Table 7 results. Table 8 shows the results of the conductivity and ppm of five aqueous Hg₂²⁺ solution prepared dissolved 0.01, 0.03, 0.05, 0.07 and 0.1 grams of HgCl₂ in 10 ml of deionizer water, each one. In a similar process were prepared five aqueous solutions EDTA-Hg₂²⁺ dissolved 0.01, 0.03, 0.05, 0.07 and 0.1 grams of HgCl₂ in 10 ml of aqueous EDTA solution and in the same way were prepared five textured soya extract-Hg₂²⁺ dissolved 0.01, 0.03, 0.05, 0.07 and 0.1 grams of HgCl₂ in 10 ml of textured soya extract.

The conductivity and ppm of the mixture of EDTA and extract of soybeans with aqueous Hg₂²⁺ solution, shown in Table VII, is a result of subtracting the conductivity or ppm of the mixtures from the EDTA and textured soya extract with HgCl₂.

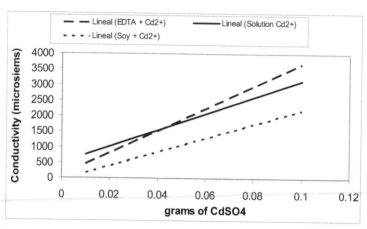

Fig. 7. Conductivity of the aqueous solution of: Cd²⁺, EDTA-Cd²⁺ and textured soya extract-Cd²⁺.

Grams HgCl₂	Hg₂²⁺ aqueous Solutions ppm (µs)		Textured soya extract-Hg₂²⁺ Ppm (µs)		EDTA-Hg₂²⁺ (5x10⁻⁴M) Ppm (µs)	
0.01	25	39.41	46	5	113	171
0.03	26.92	42.28	58	74	107.2	163.7
0.05	33.82	53.24	78	94	100	151.7
0.07	37.56	59.09	79	100	107	160.7
0.1	48.58	76.74	100	134	113.7	173

Table 8. Conductivity and ppm of aqueous solutions: Hg_2^{2+}, EDTA- Hg_2^{2+} and textured soya extract-Hg_2^{2+}.

Figure 8 is a graphic representation of Table 8 results.
Table 9 shows the results of the conductivity and ppm of five aqueous Ni^{2+} solution prepared dissolved 0.01, 0.03, 0.05, 0.07 and 0.1 grams of $NiCl_2$ in 10 ml of deionizer water, each one. In a similar process were prepared five aqueous solutions EDTA-Ni^{2+} dissolved 0.01, 0.03, 0.05, 0.07 and 0.1 grams of $NiCl_2$ in 10 ml of aqueous EDTA solution and in the same way were prepared five aqueous textured soya extract-Ni^{2+} dissolved 0.01, 0.03, 0.05, 0.07 and 0.1 grams of $NiCl_2$ in 10 ml of textured soya extract.

Grams NiCl₂	Ni²⁺ aqueous solution ppm (µs)		Textured soya extract-Ni²⁺solution Ppm (µs)		EDTA-Ni²⁺ (5x10⁻⁴M) ppm (µs)	
0.01	720.25	1052.34	210	273	450.5	716.7
0.03	1849.35	2553.34	1108	1404	997	1413.7
0.05	2677.35	3585.34	1845	2306	2653	3519.7
0.07	4093.35	5287.34	3225	3940	4088	5194.7
0.1	6069.35	7575.34	4658	5599	5607	7006.7

Table 9. Conductivity and ppm of aqueous solutions: Ni^{2+}, EDTA- Ni^{2+} and textured soya extract-Ni^{2+}.

A Comparative Study of the Chelating Effect Between Textured Soya Aqueous Extract and EDTA on Fe³⁺, Pb²⁺, Hg²⁺, Cd²⁺ and Ni²⁺ Ions

39

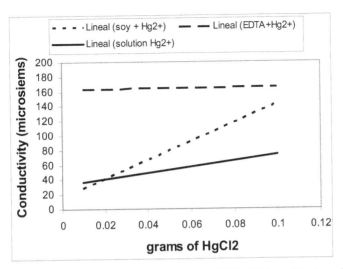

Fig. 8. Conductivity of the aqueous solution of: Hg_2^{2+}, EDTA-Hg^{2+} and textured soya extract-Hg^{2+}.

The conductivity and ppm of the mixture of EDTA and textured soya extract with aqueous Ni^{2+} solution, shown in the Table VIII, is a result of subtracting the conductivity or ppm of the mixtures from the EDTA and textured soya extract with $NiCl_2$.
Figure 9 is a graphic representation of Table 9 results.

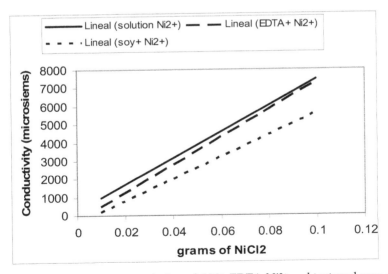

Fig. 9. Conductivity of the aqueous solution of: Ni^{2+}, EDTA-Ni^{2+} and textured soya extract-Ni^{2+}.

In Table 10 we can see the amount of metal ion sequestering for the textured soya extract in five different amounts of each salt: 0.01, 0.03, 0.05, 0.07 and 0.1 grams.

	0.01 grams	0.03 grams	0.05 grams	0.07 grams	0.1 grams
Pb^{2+}	0.0076	0.008	0.019	0.018	0.027
Fe^{3+}	0.0087	0.017	0.016	0.0089	0.021
Cd^{2+}	0.0082	0.013	0.020	0.017	0.025
Hg^{2+}	-----	-----	-----	-----	-----
Ni^{2+}	0.0075	0.012	0.018	0.014	0.027

Table 10. Amount of metal ion sequestering using textured soya extract

Figure 10 is a graphic representation of Table 10 results.

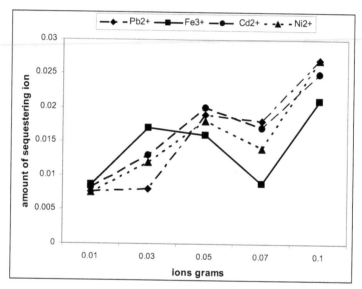

Fig. 10. Amount of metal ion sequestering using textured soya extract.

In Table 11 we can see the amount of metal ion sequestering for the EDTA in five different amounts of each salt: 0.01, 0.03, 0.05, 0.07 and 0.1 grams.

	0.01 grams	0.03 grams	0.05 grams	0.07 grams	0.1 grams
Pb^{2+}	0.0025	0.013	0.008	0.006	0.004
Fe^{3+}	0.0005	0.003	0.045	0.001	0.006
Cd^{2+}	0.0026	0.013	----	----	----
Hg^{2+}	----	----	----	----	----
Ni^{2+}	0.0032	0.013	0.001	0.0085	0.008

Table 11. Amount of metal ion sequestering using EDTA

Figure 11 is a graphic representation of Table 11 results
From the results obtained in Table 10 and Table 11, we can say that the amount of metal ion chelating increases with the increase of the concentration but the amount of salt chelated with textured soya extract is considerable major in comparison to the EDTA. In the case of Hg_2^{2+} ions, the textured soya extract and EDTA is not effective as a chelating agent. Another

A Comparative Study of the Chelating Effect Between Textured Soya Aqueous Extract and EDTA on Fe³⁺, Pb²⁺, Hg²⁺, Cd²⁺ and Ni²⁺ Ions

41

difference is that EDTA is effective as chelating agent of Cd^{2+} only with low concentrations (less to 0.04 g of $CdSO_4$ in 10 ml of EDTA solution 5×10^{-4}M) but without exception the textured soya extract is a good chelating agent.

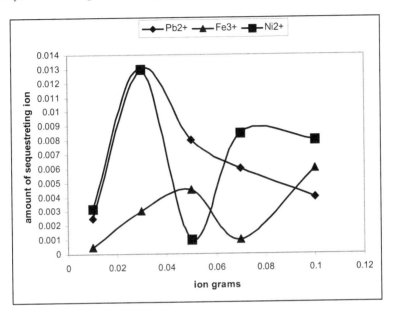

Fig. 11. Amount of metal ion sequestering using EDTA

In the case of $HgCl_2$ it is a weak electrolyte, the ionization is partial, as showing in the reaction 7.

$$HgCl_2 \ (s) \ = \ HgCl^+ \ (aq) \ + Cl^- \ (aq) \tag{7}$$

Maybe for this reason the EDTA and textured soya extract are not effective as chelating agents. But the textured soya extract is effective as chelating agent at low concentrations; nevertheless, the EDTA is not an effective chelating agent even in low concentrations.

From the results of Table 11, we are able to calculate the equilibrium constant for the textured soya extract, using the equation 6, and the results being shown in Tables 12.

	0.01 grams Textured soya extract	0.03 grams Textured soya extract	0.05 grams Textured soya extract	0.07 grams Textured Soya extract	0.1 grams Textured soya extract
Pb²⁺	0.76	0.11	0.38	0.25	0.27
Fe³⁺	0.87	0.56	0.32	0.12	0.21
Cd²⁺	0.82	0.43	0.4	0.24	0.25
Hg₂²⁺	------	----	------	----	------
Ni²⁺	0.75	0.4	0.36	0.2	0.27

Table 12. Equilibrium constant of metal ion sequestering

From the results of Table 11, we are able to calculate the equilibrium constant for the EDTA, using the equation 6, and the results being shown in Tables 13.

	0.01 grams EDTA	0.03 grams EDTA	0.05 grams EDTA	0.07 grams EDTA	0.1 grams EDTA
Pb^{2+}	0.25	0.2	0.16	0.08	0.04
Fe^{3+}	0.05	0.1	0.09	0.014	0.06
Cd^{2+}	0.26	0.43	------	----	-----
Hg_2^{2+}	------	----	------	----	-----
Ni^{2+}	0.32	0.43	0.02	0.12	0.08

Table 13. Equilibrium constant of metal ion sequestering

5. Conclusions

In the case of the ion Pb^{2+}, it can be seen that the solution of the complex EDTA with the ion P^{2+} gives a line which is very close to the reference line of the ionic solution Pb^{2+}. This indicates that the amount of Pb^{2+} ion chelated is small in comparison to the solution of the chelate formed from the textured soya extract and the Pb^{2+} ion which has a line that is way below the reference line and the EDTA.

A similar conclusion for the study with the Fe^{3+} ion can be given.

With respect to the Cd^{2+} ion, the EDTA only acts as a chelate in concentrations lower than 0.04 grams of $CdSO_4$ per 10ml of deionized water. On the other hand, the textured soya extract is a good chelate in a wider concentration range (between 0.01 and 0.1 grams of $CdSO_4$ per 10ml of deionized water).

The chelate solution of EDTA for the Hg_2^{2+} ion does not have any effect on the Hg_2^{2+} ion in the test range from 0.01 up to 0.1 grams of $HgCl_2$ per 10ml of water. However, the textured soya extracts act as a chelate only in concentrations lower than 0.15 grams of $HgCl_2$ per 10ml of deionized water. The problem presented by this salt rests on the fact that it is a weak electrolyte and when it is placed in the water, it decomposes into two ions. Since the conductivity of the solution is measured in this study, the formation of two ions has a negative effect on the measurements obtained.

Just as in the case of Pb^{2+}, Fe^{3+} and Ni^{2+} ions, the textured soya extract is a much better chelate than EDTA.

There is normally a low concentration (parts per million) of heavy ions in food. Thus our proposal of using the textured soya extract as the chelate for heavy ions instead of EDTA. In addition, as a result of the low concentration of ions, a solution with a low concentration of the textured soya extract will be used in order not to change the color, scent or taste of the food.

A problem to be considered in this application is that food is prepared with water which has salts that are ionized, and the textured soya extract will chelate some of these ions also. It will be necessary to perform tests on the food sample to determine if the application will be practical or not.

Another application possible is the extraction of heavy metals in water, cosmetics, and soils employing textured soya extract as the chelate for heavy ions instead of EDTA.

A Comparative Study of the Chelating Effect Between Textured Soya Aqueous Extract and EDTA on Fe³⁺, Pb²⁺, Hg²⁺, Cd²⁺ and Ni²⁺ Ions

43

6. Acknowledgments

This work was supported by CONACYT. Morales Elpidio, Guajardo Jesús, Lopez Francisco, Quintero Cristina, Compean Martha, Noriega María-Eugenia, would like to thank CONACYT for the grant of scholarships.

7. References

[1] Mulligan, C.N.; Young, R.N.; Gibbs, B. *Eng. Geol.* 2001, *60*, 193-207.

[2] Peters, R.W. J. Hazard. *Mater.* 1999, *66*, 151-210.

[3] Tandy Susan, Bossart Karin, Mueller Roland, Ritschel Jens, Hauser Lukas, Sculin Rainer, and Nowack Bernd, *Environ. Sci. Technol.* 2004, *38*, 937-944.

[4] Elliot, H. A.; Brown, G.A. *Water, Air, Soil Pollut.* 1989, *45*, 361-369.

[5] Lanigan RS and Yamarik TA (2002). "Final report on the safety assessment of EDTA, calcium disodium EDTA, diammonium EDTA, dipotassium EDTA, disodium EDTA, TEA-EDTA, tetrasodium EDTA, tripotassium EDTA, trisodium EDTA, HEDTA, and trisodium HEDTA". Int J Toxicol. 21 Suppl 2: 95–142.

[6] Steel, M. C; Pichtel, J. J. *Environ.Eng.* 1998, *124,* 639-645.]

[7] Pichtel, J. ; Vine, B.; Kuula-Vaisanen, P. ; Niskanen, P. *Environ. Eng. Sci.* 2001, *18,* 91-98

[8] Pichtel, J.; Pichtel, T.M. *Environ. Eng. Sci.*1997, *14*, 97-104.

[19] Kim, C.; Lee, Y.; Ong, S.K. *Chemosphere* 2003, *51*, 845-85.

[10] Ghestem, J.; Bermond, A. *Environ. Thecnol.* 1998, *19*, 409-416.

[11] Reed, B.E.; Carriere,P.C.; Moore, R. *J. Environ. Eng.* 1996, *122*, 48-50.

[12] Cline, S.R.; Reed,B.E. *J. Environ. Eng.* 1995, *121*, 700-705

[13] Van Benschoten, J.E.;Matsumoto,M.R.; Young,W.H.*J.Environ. Eng.* 1997, *127*, 217-224.

[14] Pichtel,J.; Vine,B.; Kuula-Vaisanen, P.; Niskanen, P. *Environ. Eng. Sci.* 2001, *18*, 91-98.

[15] Papassiopi, N.; Tambouris, S. Kontopoulos, A. *Water, Air, Soil, Pollut.* 1999, *109*, 1-15

[16] Xie, T.; Marshall, W.D. *J. Environ. Monit.* 2001, *3*, 411-416.

[17] Liu, Ke Shun (1997-05-01), Soybeans: *Chemistry technology and Utilization. Springer.* 532

[18] IUPAC definition of chelation.

[19] Morgan, Gilbert T.; Drew, Harry D. K. (1920).

[20] U. Kramer, JD Cotter-Howells, JM Charnock, A HJM Baker, J A C Smith. *Nature,*1996, 379: 635-638.

[21] Jurandir Vieira Magalhaes, *Proc. Natl. Acad. Sci. USA*, 2006, 103 (26):9749.

[22] Suk-Bomg Há, Aaron P. Smith, Ross Howden, Wendy M. Dietrich, Sarah Bugg, Matthew J. O'Connell, Peter B. Goldsbrough, and Chistopher S. Cobbe, *Plant Cell*, 1999, 11 (6): 1153-1164.

[23] S. J. Lippard, J. M. Berg, "Principles of Bioinorganic Chemistry" University Science Books: Mill Valley, CA; 1994 ISBN 0-935702-73-3.

[24] Michael Pidwirny, physical geography.net/fundamentals/10r

[25] Prasad (ed). Metals in the Environment. University of Hyderabad. Dekker, New York, 2001.

[26] Doja A, Roberts W., Can J. Neurol. Sci., 2006, 33 (4): 341-346.

[27] U. S. Center for Disease Control, "Death Associated with Hypocalcemia from Chelation Therapy" ,2006

[28] The 2009 Annual Report of the North Carolina Soybean Producers Association, realeased Jannary 14, 2010.

[29] Riaz MN, *Soy applications in food*. 2006. CRC Press. Pp 155-84, ISBN 0-8493-2981-7.

[30] Foote R, *Food preparation an cooking*, 1996, Cheltenham. Stanley Thornes, pp. 393. ISBN 0-7487-2566-0.

[31] Hoogenkamp, Henk W., 2005, Wallingford, Oxon, UK: CABI Pub. ISBN 0-85199-864-X.

[32] Joseph G. Endres, Soy Protein Products, 2001, AOCS Publishing. ISBN 1-893997-27-8.

3

Effect of Row-Spacing and Planting Density on Podding and Yield Performance of Early Soybean Cultivar 'Enrei' with Reference to Raceme Order

Kuniyuki Saitoh
Okayama University
Japan

1. Introduction

In Japan, the genetically modified herbicide-tolerant soybean cultivar cannot be grown in the commercial field without permission due to the public concern about the effects on the ecosystem and human health. Recently, interest for no-tilling, narrow row-spacing and dense cultivation in soybean has been increasing as a labour-saving technique. The no-tilling cultivation has an advantage in saving labor and drainage of soil, but the merit of narrow row and dense planting has not been clarified. The dense planting increases the competition among plants from the early stage and the risk of excessive growth which results in lodging. On condition that the planting density is equal, narrow row-spacing decrease the competition with plants during the earlier growth stage than wide row-spacing, and result in rapid leaf area expansion, higher crop growth rate and higher seed yield due to the development of branches, increase in the node number and pod number per node (Cooper 1977, Costa et al. 1980, Duncan 1986, Miura and Gemma 1986, Miura et al. 1987, Board et al. 1990a, 1990b, Bullock et al. 1998, Ikeda 2000). However, narrow row-spacing did not increase the yield (Beatty et al. 1982, Nakano 1989) and has been reported to even decrease the yield (Cooper and Nave 1974).

In this chapter, the factors affecting the increase in yield of narrow row and dense planting in soybean and yield determining process was clarified with reference to pod position (main stem/branches, raceme order). In order to analyze the advantages and disadvantages of narrow row and dense planting, we examined the effects of planting pattern and density on solar radiation utilization, dry-matter production and emergence of weeds.

2. Materials and methods

2.1 Plant cultivation and experimental plots

The field experiment was conducted at the Field Science Centre of Okayama University (34°41' N, 133°55' E, Japan) in 2001 and 2002. The texture of the soil was sandy clay and preceding crop was pumpkin. Indeterminate soybean (*Glycine max* (L.) Merr.) cv. 'Enrei' (maturity group III) was used. Two seeds were sown on 13 and 14 June in 2001 and 2002, respectively, with an 80cm (wide) and 30cm (narrow) row-spacing, and sparse (11.1 plants m^{-2}, 11.25 and 30cm plant spacing in wide and narrow row-spacing, respectively) and dense (22.2 plants m^{-2}, 5.6cm and 15cm in wide and narrow row-spacing, respectively) planting density. Each plots size was 57.6 m^2 (3.2×18.0m) with no replication. A basal fertilizar was applied at the rate of 2.1g N, 4.4g P and 10.0g K. Herbicide was applied to the soil surface to avoid weed emergence. The plants were thinned to a plant per hill when primary leaves were fully expanded. In wide row plots, soil molding was conducted by a rotary cultivator. The crop was irrigated with a water-spraying vinyl hose placed on every other row. Recommended pesticides were applied for the control of insects and diseases.

2.2 Growth and yield observation

Thirty plants were harvested from each plots, and ten standard plants were selected to examine the node number, main stem length, stem diameter, stem weight, and seed/stem weight ratio. Pods were distinguished on the position, main stem/branches and raceme order (Fig. 1.), and seeds were depodded manually, then weighed to record the data on yield and yield components.

The raceme orders were defined as follows (Torigoe et al. 1982). The terminal racemes appeared at the top of the stems, and first order racemes differentiate from the axil just above the petiole on the stem. The secondary racemes differentiate from both sides of the first order raceme and tertiary racemes differentiate from the sides of the secondary racemes. Racemes differentiating from both sides of the branch were classified as secondary racemes. The terminal and first order racemes, and those over secondary raceme will be collectively called basal raceme and lateral raceme, respectively. Some lateral racemes had compound leaves. The lodging score was recorded every week by measuring the angle of the main stem, and ranked 0 (erect), 1 (inclined 15 degrees), 2 (inclined 45 degrees), 3 (inclined 75 degrees) and 4 (inclined horizontally), then the average score was obtained.

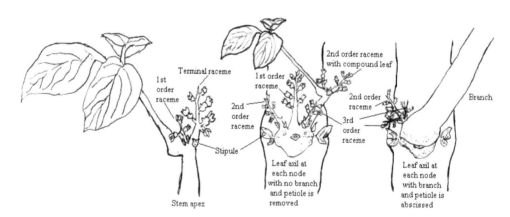

Fig. 1. Classification of raceme order in determinate type of soybean.

2.3 Dry matter production and canopy structure

Five plants (three replication for each plots) were sampled and three (nine plants for each plots) were separated into leaves, petioles, stems and pods on each main stem and branch, then measured the leaf area of a standard plant (AAM-8, Hayashidenko). Samples were air-

Effect of Row-Spacing and Planting Density on Podding and Yield Performance of Early Soybean Cultivar
'Enrei' with Reference to Raceme Order

47

dried at 80 degrees C for 48 hours and weighed. At the beginning of flowering and full seed growth stage relative PAR (photosynthetically active radiation) at each height of the canopy were measured with a long PAR sensor (LI-191S, LI-COR) in the evening under diffuse light condition. Then, canopy structures were surveyed by the stratified clip method (Monsi und Saeki 1953). From the logarithmic relationships between cumulative LAI of the canopy top and relative PAR, the canopy light extinction coefficient (k) was obtained. In addition, the relative PAR at the height of 0, 60 and 120cm above the ground was measured every 2.5 hours from 7 a.m. to 17 p.m., and diurnal change in light extinction coefficient under direct light condition was obtained.

2.4 Cumulative solar radiation within canopy

Integrated solarimeter films (R-2D, Taisei E&L) were used for the measurement of cumulative solar radiation. Film was cut in 1cm width and 2cm length, then placed at 10cm intervals on the square bars, 1cm width and 100cm length, which were installed horizontally every 15cm height from the soil surface. The dye percentages were measured every six hours by a spectro-photometer (UV-1200, Shimadzu). The dye percentages had been calibrated with the cumulated solar radiation measured by radiation sensor (LI-200SA, LI-COR). Accordingly, the distribution of solar radiation within a canopy was calculated.

2.5 Weed emergence

Three quadrats (80cm*60cm) were randomly arranged within each plots. At the beginning of flowering stage, all weeds were sampled and the number and dry-weight of each weed species were recorded.

3. Results

3.1 Growth characters

In 2001, the precipitation was 14% lower, the average mean temperature was 0.8 degree higher, and the sunshine hours was 13% longer than the normal year, and it was characterized by low rainfall, high temperature and much sunshine. In 2002, the precipitation was 56% lower, the average mean temperature was 0.9 degree higher, and the sunshine hours was 7% longer than the normal year, and it was characterized by drought, high temperature and much sunshine though lower than in 2001. The field was hit by a typhoon on Aug. 21 in 2001. There was no typhoon damage in 2002.

In both years, the number (per square meter) of nodes on the main stem, racemes with compound leaves and in total was higher, but in the number of branches was lower than in sparse plots (Table 1). The node number on the branches and in total was larger in wide plots than in narrow plots except that in sparse plots in 2001, and also that of racemes with compound leaf in 2001. The main stem length in dense plots was 2-12 cm longer than in sparse plots, and that in narrow plots was 7-16 cm shorter than in wide plots. The weight, diameter and section area of stem were larger than in sparse and narrow plots than dense and wide plots, respectively. The seed/stem weight ratio in dense plots was smaller than in sparse plots among the narrow plots, but not among the wide plots. The ratio in narrow plots was larger than in wide plots among the sparse plots, but not among the dense plots.

Year / Plot	Node number (m^{-2})				Main stem length (cm)	Stem weight (g)	Bran-ch no. (m^{-2})	Stem dia-meter (mm)	Stem section area (mm^2)	Seed / stem weight ratio
	Main stem	Bran- ch	Rac. with leaf	Total						
2001										
Wide/Sparse	150	316	137	602	63.4	18.3	66	9.4	53.0	2.20
Wide/ Dense	290	192	183	665	69.6	8.7	74	6.9	30.4	2.51
Narrow/Sparse	141	239	211	591	47.4	18.1	60	9.2	56.4	2.58
Narrow/Dense	296	342	307	944	55.7	12.2	102	7.8	37.1	2.49
LSD$_{(0.05)}$	9	ns	33	54	3.3	1.4	9	0.4	4.4	ns
2002										
Wide/Sparse	159	272	71	502	61.2	12.5	60	8.5	43.5	1.98
Wide/ Dense	301	248	121	670	63.5	7.7	89	7.0	27.8	2.06
Narrow/Sparse	162	324	89	576	53.6	13.6	70	9.1	49.4	3.39
Narrow/Dense	318	347	122	787	65.3	10.2	111	7.8	38.2	2.25
LSD$_{(0.05)}$	7	53	24	67	2.5	1.3	14	0.1	3.4	0.50

Values are means of twelve plants. 'ns' means no siginificant difference at 5% level.

Table 1. Growth characteristics (2001, 2002).

3.2 Seed yield and yield components

In both years, seed yields in dense plots and narrow plots were larger than sparse plots and wide plots, respectively, and those in 2001 were higher than in 2002 because of the much sunshine hours (Fig. 2, Table 2). The highest yield, 668 g m^{-2}, was obtained in narrow/dense plots in 2001. A close correlation (r=0.934, P<0.01) was observed between seed yield and pod

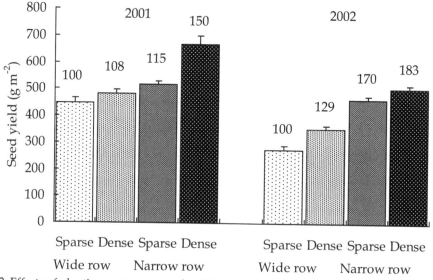

Fig. 2. Effects of planting pattern on seed yield of soybean.

Effect of Row-Spacing and Planting Density on Podding and Yield Performance of Early Soybean Cultivar
'Enrei' with Reference to Raceme Order

49

number, indicating that seed yield was determined by the pod number. Seed number per pod and seed setting ratio were not significantly different among plots, and 100 seeds weight in narrow plots tended to be slightly heavier than in wide plots, but the difference was not significant.

The pod number on the main stem relative to the total was higher in dense plots than in sparse plots, and that on the branches was higher in narrow plots than in wide plots (Table 3). The percentage share of basal raceme was higher in 2002 than in 2001. The percentage share of racemes with compound leaves was higher in dense plots than in sparse plots, and was also higher in narrow plots than in wide plots, especially in 2001.

Year / plot	Seed yield ($g\,m^{-2}$)	Pod number (m^{-2})	Seed number per pod	100 seeds weight (g)	Seed setting ratio (%)
2001					
Wide/Sparse	446	894	2.03	30.2	95.0
Wide/ Dense	483	904	1.99	31.1	96.5
Narrow/Sparse	515	1011	1.96	32.3	95.5
Narrow/Dense	668	1256	1.99	31.5	97.4
LSD$_{(0.05)}$	45	84	ns	ns	ns
2002					
Wide/Sparse	274	766	2.04	26.5	83.7
Wide/ Dense	354	893	2.01	27.5	91.3
Narrow/Sparse	464	910	2.02	33.6	87.5
Narrow/Dense	503	993	2.04	31.8	92.2
LSD$_{(0.05)}$	26	83	ns	ns	4.5

Values are means of twelve plants.
'ns' means no siginificant difference at 5% level.

Table 2. Seed yield and yield components (2001, 2002).

Year / plot	Main stem	Branch	Basal raceme	Raceme with leaf	Upper raceme	Total
2001						
Wide/Sparse	377 (42)	518 (58)	367 (41)	262 (29)	266 (30)	894 (100)
Wide/ Dense	685 (76)	219 (24)	359 (40)	321 (36)	223 (24)	904 (100)
Narrow/Sparse	384 (38)	627 (62)	333 (33)	416 (41)	262 (22)	1011 (100)
Narrow/Dense	702 (56)	553 (44)	456 (36)	524 (42)	276 (22)	1256 (100)
LSD$_{(0.05)}$	61	93	ns	66	ns	85
2002						
Wide/Sparse	337 (44)	429 (56)	446 (58)	119 (16)	201 (26)	766 (100)
Wide/ Dense	567 (63)	326 (37)	464 (52)	205 (23)	223 (25)	893 (100)
Narrow/Sparse	292 (32)	618 (68)	480 (53)	154 (17)	276 (30)	910 (100)
Narrow/Dense	607 (61)	387 (39)	536 (54)	244 (25)	213 (21)	993 (100)
LSD$_{(0.05)}$	48	72	ns	43	ns	83

Values are means of twelve plants. 'ns' means no siginificant difference at 5% level.
Values in parentheses are relative to total (100).

Table 3. Pod number on main stem or branch and raceme order (2001, 2002).

3.3 Dry weight and leaf area index

At each growth stage, the dry-weight tended to be heavier in dense plots than in sparse plots, but the difference was not significant (Fig. 3). The dry-weight tended to be heavier in narrow plots than in wide plots except that in sparse plots at 44 days after sowing (DAS) and in dense plots at 65 DAS. At 107 DAS, the dry-weight was heaviest in narrow/dense plots and became lighter in the order of wide/dense plots > narrow/sparse plots > wide/sparse plots.

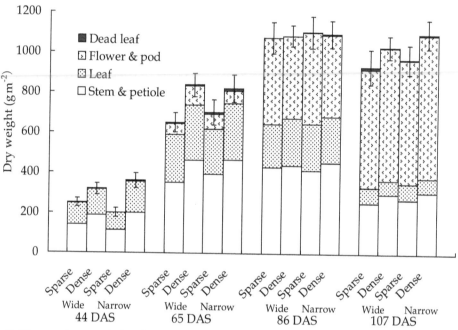

Fig. 3. Changes in cumulative dry-weight of different plant parts during growth (2001).

The leaf area index (LAI) tended to be larger in dense plots than in sparse plots, and in narrow plots than in wide plots especially at 65 DAS, when LAI in dense plots exceeded 8 (Fig. 4).

3.4 Canopy structure

At the flowering stage, the higher the canopy layer, the larger the leaf area from 20 to 100 cm above the ground in wide/dense plots, and the larger leaf area was distributed at a 40-100 cm height in narrow/sparse plots (Fig. 5). In dense plots, leaf area was concentrated in the 80-100 cm layer above the ground especially in narrow plots. The total dry-weight of non-assimilative organ was heavier in narrow plots than in wide plots. The light extinction coefficients (k), the lower value indicates that the canopy has a good light-intercepting characteristic, was in the order of narrow/dense (0.60) < wide/dense (0.68) < narrow/sparse (0.73) < wide/sparse (0.81). It was clear that the light penetrated into a deeper layer of the canopy when planted dense and narrow row-spacing. The order of k at the seed growth stage coincided with that at the flowering stage (data not shown).

Effect of Row-Spacing and Planting Density on Podding and Yield Performance of Early Soybean Cultivar
'Enrei' with Reference to Raceme Order

51

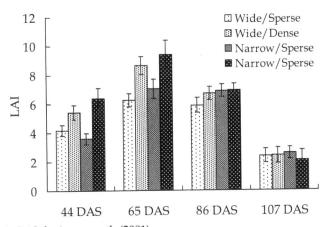

Fig. 4. Changes in LAI during growth (2001).

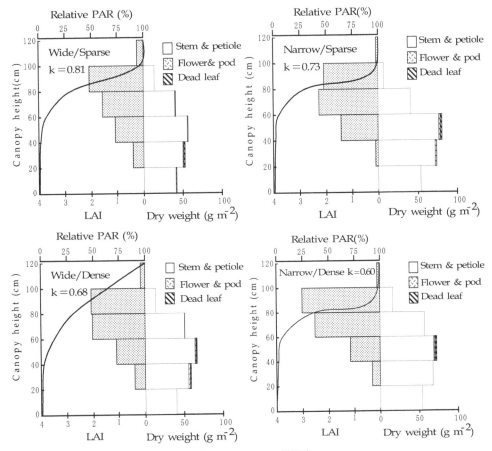

Fig. 5. Canopy structures at the full-flowering stage (2001).

3.5 Diurnal change in canopy light extinction coefficient (k)

The k-value measured under direct sunlight was higher in the morning and evening, and decreased during the daytime (Fig. 6). The k-values in the morning and evening were similar to those measured under diffuse light (Fig. 5), which were lower in dense and narrow row plots. At midday, k showed the lowest value in wide plots, which suggested that the direct sunlight reached the furrow surface in the non-closed canopy in wide row plots. The extent of variation during the daytime was small in narrow plots due to the closed canopy.

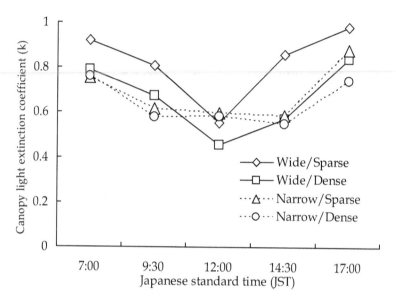

Fig. 6. Diurnal change in canopy light extinction coefficient at the beginning of the flower stage (2001).

3.6 Distribution of cumulative solar radiation at each height within canopy

The cumulative solar radiation at every height was lower in dense plots than in sparse plots, and was lower near the row (plant) and higher at the furrow in a direction perpendicular to the row (Fig. 7). In narrow row plots, the cumulative solar radiation was lower in dense plots than in sparse plots, and the difference between that on the row and furrow was small.

3.7 Changes in lodging score

In 2002, lodging did not occur in any plot. In 2001, the lodging score increased in narrow/sparse plots at 34 DAS due to a rainstorm, followed by the gradual increase in wide/sparse plots, and was larger in narrow row plots than in wide row plots (Fig. 8). At 71 DAS, when a typhoon hit, the lodging score increased markedly in dense plots, and was slightly larger in narrow/dense plots than in wide/dense plots. After lodging, plants could not recover during the later growth period.

Effect of Row-Spacing and Planting Density on Podding and Yield Performance of Early Soybean Cultivar
'Enrei' with Reference to Raceme Order

53

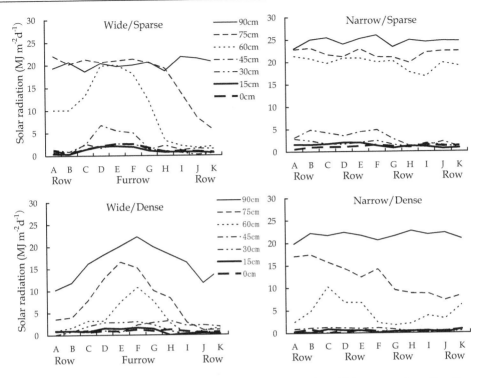

Fig. 7. Distribution of cumulative solar radiation at each height within canopy in a direction perpendicular to the row at the beginning flower stage (2001).

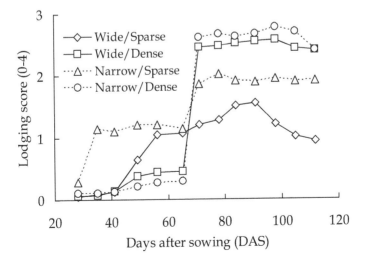

Fig. 8. Changes in lodging score (2001).

3.8 Weed emergence

More weed plants appeared in 2002 than in 2001. *Portulaca* and *Cyperus* species were dominant in 2001, and *Digitaria* and *Galinsoga* in 2002. In both years, there were fewer emerged weeds in narrow plots than in wide plots.

Year/Plot	*Amaranthus viridis*	*Portulaca oleracea*	*Digitaria ciliaris*	*Cyperus*	*Rorippa indica*	*Galinsoga ciliata*	*Setaria viridi*	*Chenopodium album*	*Euphorbia supina*	*Mollugo pentaphylla*	Total
2001											
Wide/Sparse	7.6	9.7	-	8.3	2.1	2.1	2.1	2.1	-	-	34.0
Wide/ Dense	5.2	16.0	-	7.6	2.1	2.1	2.1	9.0	-	3.1	47.2
Narrow/Sparse	-	2.1	-	-	-	-	-	-	-	-	2.1
Narrow/Dense	-	14.6	-	-	-	2.1	2.1	2.1	-	-	20.8
2002											
Wide/Sparse	-	52.8	11.1	-	-	60.2	-	-	28.7	-	124.1
Wide/ Dense	-	-	23.1	-	-	94.4	-	-	-	-	117.6
Narrow/Sparse	-	63.9	9.3	-	-	11.1	3.7	-	25.0	-	78.7
Narrow/Dense	-	-	18.5	-	-	45.4	4.6	-	-	-	68.5

Values indicate the number of weed plants. Average of three quadrats (80cm * 60cm) .

Table 4. Emergence of weeds at the beginning of flowering of soybean.

4. Discussion

In soybean, dense planting has been reported to increase the node number, pod number and therefore seed yield without the consideration of lodging (Nakaseko and Goto 1975, Costa et al. 1980, Miura et al. 1987, Saitoh et al. 1998a). The square- or triangular-shape planting increased the space occupied by plants than rectangular-shape planting, and promoted the development of branches, thus increasing the seed yield (Cooper 1977, Costa et al. 1980, Duncan 1986, Miura and Gemma 1986, Miura et al. 1987, Board et al. 1990b, Ikeda 2000). Nakano et al. (2001) also reported that planting pattern affected the light environment within the canopy, which determined the branch node number, pod number and seed yield. In the present study, the seed yield was in the order of narrow/dense > narrow/sparse > wide/dense > wide/sparse (Table 2, Fig. 2), and the yield increase in narrow row planting was due to the yield increase on the branches especially on the raceme with compound leaves (Table 3).

The raceme with compound leaves is morphologically the same as a branch. The branch differentiates on the leaf axil just above the petiole on the main stem, and the raceme with compound leaves differentiates on the left and right axils of the basal raceme in the upper node of the main stem and branches, and develops a stem with one to four leaves. In a previous study, the differentiated racemes developed compound leaves when assimilates were supplied to the raceme (Saitoh et al. 2001). In the present two- year study, seed yield was positively correlated with total pod number (r=0.934, P<0.001) and pod number on racemes with compound leaves (r=0.864, P<0.01). Thus the increase in the pod number on the raceme with compound leaves contributed to the increase in seed yield.

The longer sunshine hours accelerated the source activity and increased assimilates were supplied to the axil of each node. Our three-year planting density experiment showed that the number of floral buds on racemes with compound leaves increased markedly in the year with longer sunshine hours (Saitoh et al. 1998a), and the pod number on racemes with compound leaves increased especially when the twelfth node was isolated by pruning the top above the twelfth node and removing all of the leaves, petioles and floral organs except those on the twelfth node at the flowering stage. Under such conditions, assimilates were concentrated to the twelfth node (Saitoh et al. 1998b), and the number of racemes with compound leaves on the main stem and branches increased when the leaves on branches and main stem were removed, respectively (Saitoh et al. 2001).

The present study revealed that the increase in pod number by narrow row planting was due to the increase in that on the racemes with compound leaves suggesting that the microclimate within canopy affected the development of racemes with compound leaves in narrow row-spacing. The narrow row-spacing canopy had a lower light extinction coefficient, i.e., better light-intercepting characteristics (Fig. 5).

In wide row-spacing, solar radiation was distributed non-uniformly, penetrated a deeper layer of the canopy due to fewer leaves distributed within the furrow, and decreased markedly above the row space (Fig. 7). In narrow row-spacing, solar radiation was distributed uniformly, the difference between the row and furrow was small, so that many racemes developed compound leaves due to the surplus assimilative supply to the raceme from the upper layer of canopy. The raceme with compound leaves is not only a sink organ, but also a source organ.

The canopy light extinction coefficients (k) measured under direct sunlight decreased during the daytime (Fig. 6). The decrease in k-value means that the sunlight penetrated uniformly into a deeper layer of the closed canopy with a higher LAI, however, sunlight reached a deeper layer directly and leaves received the excess light in non-closed canopy with lower LAI like wide row-spacing. This suggests that the k-value during the daytime can not evaluate the light intercepting characteristics in non-uniformly foliage distributed canopy.

The comparison of dry matter production in the plants with different planting patterns revealed that dry-weight was heavier and LAI was larger in dense plots than in sparse plots along as shown by others (Shibles and Weber 1965, Sugiyama et al. 1967, Asanuma et al. 1977, and also in narrow row-spacing than wide row-spacing (Fig. 3, 4) in accordance with the previous studies (Bullokck et al. 1998, Duncan 1986,Shibles and Weber 1965, 1966). In narrow row-spacing, the distance between plants was longer than in wide row-spacing, so that the canopy had a better light-intercepting environment, which accelerated the development of branches and racemes with compound leaves and the expansion of leaf area during the earlier stage, though, LAI in dense planting at 65 DAS exceeded 8, which means over luxuriant growth (Sugiyama et al. 1967).

Next, we should consider the effects of lodging. The lodging score was larger in narrow plots than in wide plots, (Fig. 8). This is because the distance between plants was longer in narrow row-spacing, and there was less mutual support with the neighboring plants. After the full flowering stage, a large amount of foliage was distributed in the upper layer of the canopy in the narrow/sparse plots (Fig. 5), and the higher the center of gravity, the higher the susceptibility to lodging. In narrow row-spacing, the main stem length was 15cm shorter and 0.9mm thicker than in wide row-spacing in 2001 (Table 1) because the competition between plants for elongation growth decreased due to the longer distance between plants. Despite this, the lodging score was larger in narrow row-spacing, meaning that the lodging of soybean was influenced by the above ground weight and center of gravity than the main stem length and stem thickness. Further study is needed to analyze the factors affecting the lodging tolerance in soybean.

Finally, let me consider about the weed management. In narrow row-spacing, we should eradicate weeds by hand if early weed control fails. It is impossible to kill weeds by cultivator after sowing. It was already demonstrated that the narrow row cultivation decreased weeds emergence and the alternative application of herbicide to soil or foliage (Gramineae weeds) could control weeds with labour saving and stability (Ohdan et al. 2005). Present results also showed that the less number of weeds were appeared in narrow row plots than in wide row plots (Table 4), in both plots herbicide was applied to the soil surface after sowing and the soil molding was conducted with a rotary cultivator in wide row plots. The dry-weight of weeds per square-meter was about 2g, which was extremely less than that of soybean, 300-400 g m^{-2}, i.e., weeds could be controlled sufficiently. We considered that weeds could be controlled by one application of herbicide to the soil surface after sowing. If we failed to kill weeds by the soil applied herbicide, the additional application of bentazone, newly registered foliar applied herbicide in Japan, can be used after sowing.

5. Conclusion

The narrow row-spacing (wide distance between plants) and dense planting in soybean increase seed yield than in the wide row-spacing (narrow distance between plants), which was caused by the decrease in competition among plants for elongation growth, the promotion of branch development, the development of racemes with compound leaves, and the increase in pod number due to the uniform light environment within the upper layer of canopy. The improvement of lodging tolerance and perfect weed control will be needed in the narrow row and dense planting of soybean were considered to be needed.

6. References

Asanuma, K., Naka, J. & Kogure, K. (1977). On the relation between dry matter production and plant density in autumn type soybeans. *Technical Bulletin Faculty of Agriculture Kagawa University*, Vol.28, pp. 11-18, ISSN 0368-5128.

Beatty, K.D., Eldridge, I.L. & Simpson, A.M.Jr. (1982). Soybean response to different planting patterns and dates, *Agronm Journal*, Vol.74, pp. 859-862, ISSN 0002-1962.

Bullokck, D., Khan, S. & Rayburn, A. (1998) Soybean yield response to narrow rows is largely due to enhanced early growth, *Crop Science*, Vol.38, pp. 1011-1016, ISSN 0011-183X.

Board, J.E., Harville, B.G. & Saxton A.M. (1990a). Narrow-row seed-yield enhancement in determinate soybean. *Agronm Journal*, Vol.82, pp. 64-68, ISSN 0002-1962.

Board, J.E., Harville, B.G. & Saxton, A.M. (1990b). Branch dry weight in relation to yield increases in narrow-row soybean, *Agronm Journal*, Vol.82, pp. 540-544, ISSN 0002-1962.

Cooper, R.L. (1971). Influence of soybean production practices on lodging and seed yield in highly productive environments, *Agronm Journal*, Vol.63, pp. 490-493, ISSN 0002-1962.

Cooper, R.L. & Nave, W.R. (1974). Effect of plant population and row width on soybean yield and harvesting loss, Transactions of the ASAE, Vol.17, pp. 801-805. ISSN 0001-2351

Cooper, R.L. (1977). Response of soybean cultivars to narrow rows and planting rates under weed-free conditions, *Agronm Journal*, Vol.69, pp. 89-92, ISSN 0002-1962.

Effect of Row-Spacing and Planting Density on Podding and Yield Performance of Early Soybean Cultivar
'Enrei' with Reference to Raceme Order

57

Costa, J.A., Oplinger, E.S. & Pendleton, J.W. (1980). Response of soybean cultivars to planting patterns, *Agronm Journal*, Vol.72, pp. 153-156, ISSN 0002-1962.

Duncan, W.G. (1986). Planting patterns and soybean yield, *Crop Science*, Vol.26, pp. 584-588, ISSN 0011-183X.

Ikeda, T. (2000). Some factors related with net production of soybean population, *Japanese Journal of Crop Science*, Vol.69, pp. 12-19, ISSN 0011-1848.

Miura, H. & Gemma, T. (1986). Effect of square planting on yield and its components of soybean under different levels of planting density, *Japanese Journal of Crop Science*, Vol.55, pp. 483-488, ISSN 0011-1848.

Miura, H., Wijeyathungam , K. & Gemma T. (1987). Variation in seed yield of soybean as affected by planting patterns, *Japanese Journal of Crop Science*, Vol.56, pp. 652-656, ISSN 0011-1848.

Monsi, M. & Saeki, T. (1953). Über den Lichtfactor in den Pflanzengesell schaften und seine Bedeutung für die Stoffproduktion, *Japanese Journal of Botany*,. Vol.14, pp. 22-52.

Nakano, H. (1989). Effect of narrow spacing on grain yield in soybean, *Japanese Journal of Crop Science*, Vol.58, pp. 133-134, ISSN 0011-1848.

Nakano, H., Komoto, K. & Ishida, K. (2001). Effect of planting pattern on development and growth of the branch from each node on the main stem in soybean, *Japanese Journal of Crop Science*, Vol.70, pp. 40-46, ISSN 0011-1848.

Nakaseko, K. & Goto, K. (1975). Comparative studies on dry matter production, plant type and productivity in leguminus crop. III. Dry matter production of soybean plant at various population densities, *Japanese Journal of Crop Science*, Vol.44 (Extra issue 2), pp. 71-72, ISSN 0011-1848.

Ohdan, H., Sumiyoshi, T. & Koarai, A. 2005. Efficient weed control using herbicides and narrow row-dense planting for cultivation with no-intertillage of soybean cv. 'Sachiyutaka'. *Report Kyushu Branch Crop Science Society of Japan*, Vol.71, 30-32, ISSN 0285-3507.

Saitoh, K., Isobe, S. & Kuroda T. (1998a). Significance of flower differentiation and development in the process of determining soybean yield : Relation between the number of pods and flowers, *Japanese Journal of Crop Science*, Vol.67, pp. 70-78, ISSN 0011-1848.

Saitoh, K., Isobe, S. & Kuroda T. (1998b). Pod elongation and seed growth as influenced by nodal position on stem and raceme order in a determinate type of soybean, *Japanese Journal of Crop Science*, Vol.67, pp. 85-90, ISSN 0011-1848.

Saitoh, K., Isobe, S., Seguchi, Y. & Kuroda T. (2001). Effects of source and/or sink restriction on the number of flowers, yield and dry-matter production in field-grown soybean, *Japanese Journal of Crop Science*, Vol.70, pp. 365-372, ISSN 0011-1848.

Shibles, R.M., & Weber. C.R. (1965). Leaf area, solar radiation and dry matter production by soybeans, *Crop Science*, Vol.5, pp. 575-578, ISSN 0011-183X.

Shibles, R.M., & Weber. C.R. (1966). Interception of solar radiation and dry matter production by various soybean planting patterns *Crop Science*, Vol.6, pp. 55-59, ISSN 0011-183X.

Sugiyama, S., Matsuzawa, H., Horiuch, T. &. Kawashima, R. (1967) Effect of various planting density and pattern on growth and seed yield of soybean, Bulletin of the Nagano Agricultural Experiment Station, Vol.32, pp.34-35, ISSN 0388838X.

Torigoe Y., Shinji H. & Kurihara H. (1982). Studies on developmental morphology and yield determining process of soybeans. II. Developmental regularity of flower clusters and flowering habit from a view point of gross morphology, *Japanese Journal of Crop Science*, Vol.51, pp. 89-96, ISSN 0011-1848.

4

Effect of Nitrate on Nodulation and Nitrogen Fixation of Soybean

Takuji Ohyama[1,2] et al.*
1Faculty of Agriculture, Niigata University
2Quantum Beam Science Directorate, Japan Atomic Energy Agency
Japan

1. Introduction

1.1 Biological nitrogen fixation and nitrogen nutrition in soybean plants

Biological nitrogen fixation is one of the most important processes for ecosystem to access available N for all living organisms. Although N_2 consists 78% of atmosphere, but the triple bond between two N atoms is very stable, and only a few group of prokaryotes can fix N_2 to ammonia by the enzyme nitrogenase. Annual rate of natural nitrogen fixation is estimated about 232×10^6 t, and the 97% depends on biological nitrogen fixation (Bloom, 2011). This exceeds the rate of chemical nitrogen fertilizer uses about 100×10^6 t in 2009. Soybean can use N_2, though symbiosis with nitrogen fixing soil bacteria, rhizobia, to make root nodules for harboring them.

Soybean (*Glycine max* [L.] Merr.) is a major grain legume crop for feeding humans and livestock. It serves as an important oil and protein source for large population residing in Asia and the American continents. The current global soybean production was 231×10^6 t in 2008 (FAOSTAT). It is a crop predominantly cultivated in U.S.A., Brazil, Argentina and China, which together contribute nearly 87 percent of the total world produce in 2008. Soybean has become the raw materials for diversity of agricultural and industrial uses.

Soybean seeds contain a high proportion of protein, about 40% based on dry weight, therefore, they require a large amount of nitrogen to get a high yield. About 8 kg N is required for 100 kg of soybean seed production. Soybean can use atmospheric dinitrogen (N_2) by nitrogen fixation of root nodules associated with soil bacteria, rhizobia. Soybean plants can absorb combined nitrogen such as nitrate for their nutrition either from soil mineralized N or fertilizer N.

It is well known that heavy supply of nitrogen fertilizer often causes the inhibition of nodulation and nitrogen fixation. Therefore, only a little or no nitrogen fertilizer is

* Hiroyuki Fujikake[1], Hiroyuki Yashima[1], Sayuri Tanabata[3], Shinji Ishikawa[1], Takashi Sato[4], Toshikazu Nishiwaki[5], Norikuni Ohtake[1], Kuni Sueyoshi[1], Satomi Ishii[2] and Shu Fujimaki[2]
[1] Faculty of Agriculture, Niigata University,
[2] Quantum Beam Science Directorate, Japan Atomic Energy Agency, [3]Agricultural Research Institute, Ibaraki Agricultural Center, [4]Faculty of Bioresource Sciences, Akita Prefectural University, [5] Food Research Center, Niigata Agricultural Research Institute, Japan

practically applied for soybean production. However, soybean plants only depend on the nitrogen fixation shows poor growth and low seed yield, because of the early decline in photosynthesis by decreased supply of nitrogen during the pod filling stage. Harper (1974) reported that both soil N and symbiotic N are required for the optimum soybean production.

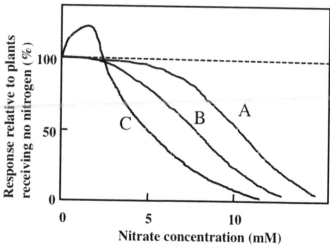

A: nodule number per a plant, B: Nitrogen fixation activity per g dry weight of nodules, C: Nodule mass per a plant.

Fig. 1. Response of legume nodules to nitrate proposed by Streeter.

The inhibitory effect of nitate on nodulaion was early reported by Fred and Graul (1916) as cited in Streeter (1988), however, the precise mechanism for the inhibition of nodulation and nitrogen fixation has not been fully understood. In the review article for inhibition of legume nodule formation and N_2 fixation by nitrate written by Streeter (1988), he proposed the responses to nitrate illustrated in Fig. 1. Curve A represents nodule number per a plant, which appears a relatively high nitrate concentration. Curve B is on nitrogen fixation activity per unit mass (g dry weight) of nodules. Curve C shows the growth response (nodule mass per a plant), this response is most sensitive to nitrate concentration, although a low concentration of nitrate as low as about 1-2 mM nitrate promotes nodule growth.

1.2 Nodule structure and function of soybean

Soybean nodule appears about 10 days after sowing when inoculated with compatible strain of rhizobia, and it grows about 3mm until about 20 days after planting (Fig. 2. A.) . The nodules start to fix nitrogen (Sato et al., 2001, Ito et al., 2006). The maximum size of nodule reaches maximum about 6-7 mm diameter, and then they eventually senesce and degrade. Soybean nodule is classified to a determinate type nodule, which has a spherical form, and nodule growth is mainly due to cell expansion after initial cell proliferation and development.

Fig. 2.B shows the structural model of a soybean nodule attached to the root. The soybean nodule has the symbiotic region (or infected region in synonym) in the center, which

A   **B**

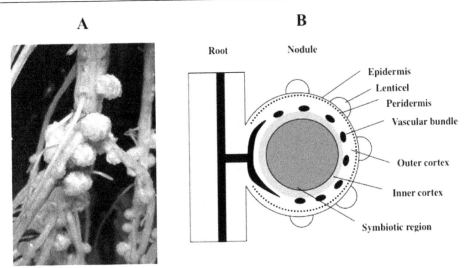

A: Photograph of soybean nodules formed in the roots of a plant cultivated by hydroponics. B: Structural model of soybean nodule.

Fig. 2. Soybean root nodules.

consists the mosaic of large infected cells and small uninfected cells. The infected cells are filled with bacteroids (the symbiotic forms of rhizobia) and they are easily recognized by the red color with nodule specific protein, leghemoglobin (Lb). The nitrogenase, an enzyme to fix N_2 in bacteroid, is very susceptible to free O_2 and irreversibly destroyed by O_2, therefore, free O_2 concentration should be kept very low in symbiotic region of nodules. There are four major components of Lb, Lba, Lbc1, Lbc2, and Lbc3 (Sato et al., 1998, 1999a). The Lb in legume nodules solves the dilemma to keep free O_2 concentration low and sufficient supply of O_2 for bacteroid respiration to support nitrogen fixation and the assimilation. Lb is a most abundant protein in nodules (about 20% of total protein) and it can bind with O_2 to form LbO_2 to decrease free O_2 concentration in the infected cells. On the other hand, nitrogen fixation and assimilation processes require a large amount of energy and reductant produced by O_2 respiration, therefore, nodule respiration is about four times higher than that of roots based on dry weight. To support active respiration, abundant supply of O_2 is essential.

Symbiotic region is surrounded by nodule cortex where the network of vascular bundles surrounding the symbiotic region to supply photoassimilate to bacteroids and to receive N_2 fixation products from them. Nodule cortex consists of inner cortex with small cells and outer cortex with large loosely packed cells. The sclerenchyma cells, which have thick cell wall were located in the outer cortex. O_2 concentration decreases sharply through the inner cortex, and the O_2 permeability is flexibly controlled by this layer (Witty and Minchin 1990, Hunt and Layzell 1993). It is hypothesized that a reversible exchange of intercellular water by the inner cortical cells plays a role in the regulation of nodule conductance to O_2 diffusion (Serraj et al., 1995, 1998, Fleurat-Lessard et al., 2005). There are lenticels outside of nodules and one layer of epidermis. Under the epidermis, there is a peridermis, a tightly packed one layer of cells, which may restrict free diffusion of solutes between inside the nodule and medium solution.

The group of positron-emitting tracer imaging system (PETIS) for plant analysis in Quantum Beam Science Directorate, Japan Atomic Energy Agency, developed a novel method of non-invasive observation and quantification of nitrogen fixation in intact soybean plants (cv. Williams) with nodules using ^{13}N-labeled nitrogen gas ($[^{13}N]N_2$) tracer and a PETIS (Ishii et al., 2009, Fujimaki et al., 2010). CO_2 gas was irradiated with a proton beam delivered from a cyclotron (Takasaki Advanced Radiation Research Institute, Japan Atomic Energy Agency) to produce ^{13}N nuclei by the ^{16}O (p, α) ^{13}N nuclear reaction. $[^{13}N]N_2$ was isolated from the resulting gas using gas chromatography and then mixed with appropriate composition of oxygen and (non-radioactive) nitrogen gases for the following feeding experiment. The total time required for the purification procedures was approximately 15 min, which is about 1.5 times the half-life of ^{13}N (only 9.97 min) and short enough to yield sufficient radioactivity of the tracer.

PETIS is one of the most advanced imaging methods today, which provides serial images of movement of positron-emitting radiotracers inside living plant bodies, like a video camera. The root of an intact test plant with nodules was immersed in a hydroponic culture solution in an acrylic box sealed with plastic clay to prevent leakage of the fed gas. This set-up was placed at the midpoint between the opposing detector heads of the PETIS apparatus so that the underground part in the acrylic box was in the field of view (Fig. 3.). The tracer gas was introduced into the box and the solution level was lowered simultaneously, then this was kept for 10 min for exposure of the nodules to the tracer gas. Finally, the tracer gas was flushed out by flowing the ambient air into the box. The two-dimensional distribution of ^{13}N in the field of view was continuously monitored by PETIS for 1 h.

As a result, obvious signals of ^{13}N were observed at the positions of the nodules (Fig. 4.). Moreover, the rates of nitrogen fixation in the whole nodules were quantitatively estimated from the PETIS data. The nitrogen fixation rate of the whole nodules was estimated at 7 μg N_2 h^{-1} in this case. The largest advantage of this method is that it is non-invasive. The instant response of fixation activities to nitrate application will be examined in a future study.

Soybean nodule is highly organized complex organ as shown by the distribution of minerals examined by EPMA (Electron Probe X-ray Microanalysis) (Mizukoshi et al., 1995). Fig. 5. shows the distribution of minerals in nodulated roots. The concentrations of N and P were higher but those of K and Cl were lower in the symbiotic region compared with nodule cortex. Ca was locally distributed in the surface layer, sclerenchyma cells and inner cortex, but the content was low in the central symbiotic region. Mg specifically accumulated in the inner and outer cortex inside sclerenchyma cells but not out side them (Mizukoshi et al., 1995).

Fig. 6. shows the outline of N metabolism in soybean nodule (Ohyama et al., 2009). Ammonia is known to be the initial product of nitrogenase. After discovering a new enzyme glutamate synthase (GOGAT) in *Aerobacter aerogenes*, it is confirmed that ammonia can be assimilated via alternative of GDH, via glutamine synthetase (GS) and GOGAT pathway (Ohyama et al. 2009). From the result obtained by the $^{15}N_2$ pulse chase experiment, the ammonia fixed by nitrogenase in bacteroids is rapidly incorporated into of amido-N of glutamine, followed by glutamate, and amino-N of glutamine in this sequence was in accordance with the initial assimilatory pathway be GS/GOGAT pathway rather than GDH. This was supported by the evidence that the rapid decline of ^{15}N in glutamine but not glutamate immediately after changing to the chase period. A major part of fixed N was used for purine synthesis in infected cells then uric acid is transported to the uninfected cells, then degraded into allantoin and allantoate. All the species in Phaseoleae (soybean, common bean, cowpea etc.) and some species in Robinieae, Indigoforeae and Desmodieae transport ureides (Atkins, 1991). Reviews on ureide biosynthesis in legume nodules were published (Schubert, 1986; Tajima et al., 2004). We compared the labeling patterns of ureides and amino acids from $^{15}N_2$ and $^{15}NO_3^-$ (Ohyama & Kumazawa, 1979), and the labeling pattern indicated that most of ureides derived from fixed N rather than absorbed N.

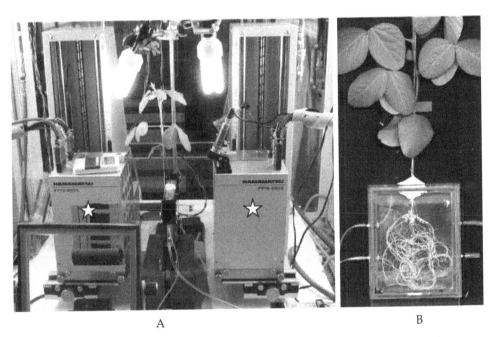

A   B

Fig. 3. Set-up for the PETIS experiment (A) and a test plant (B). Star signs indicate the opposing detector heads of the PETIS apparatus.

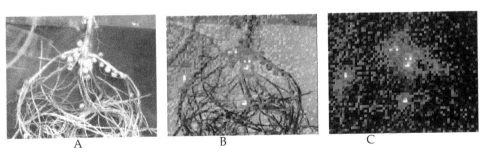

A   B   C

A: Nodulated roots of a test plant.
B: The merged image of nodulated root and radioactivity in the same view.
C: PETIS image of radioactivity.

Fig. 4. Image of radioactivity after exposure the nodulated soybean roots to $^{13}N_2$.

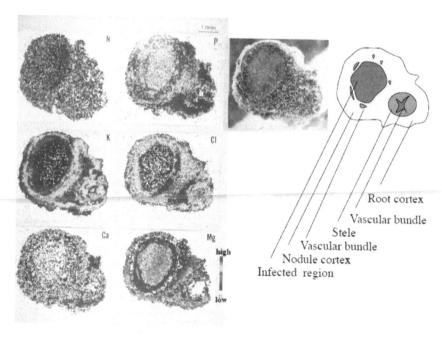

Fig. 5. Distribution of N, P, K, Cl, Ca and Mg in a nodule and root. The concentration is higher in red, orange, yellow, green, blue and white in this sequence.

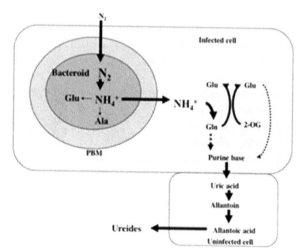

Gln: glutamine, Glu: glutamate, 2-OG: 2-oxoglutarate, PBM: peribacteroid membrane.

Fig. 6. A model of the N flow of fixed N_2 in soybean nodules.

Fig. 7. shows the model of nitrogen assimilation and transport of N derived from N_2 fixation and NO_3^- absorption in soybean plants (Ohyama &Kumazawa, 1978, 1979, 1980abc, 1981ab, 1983, 1984, Ohyama et al., 2009) . The N fixed in noudle is exported to the host plant as in the form of allantoin and allantoate about 80-90% of total N. On the other hand, some part of the NO_3^- absorbed in the roots are reduced in the roots to NO_2^- by nitrate reductase, then the NO_2^- is further reduced to NH_4^+ by plastidic nitrite reductase, then the NH_4^+ is assimilated by GS/GOGAT pathway in the roots, and mainly metabolized to asparagine then transported to shoot via xylem. Some part of NO_3^- is directly transported through xylem to the shoots and reduced in leaves. Ohtake et al. (1995) reported the seasonal changes in amino acid composition in xylem sap of soybean and they confirmed that asparagine was the principal amino acids in xylem sap collected from basal cut end of the stem at any stages.

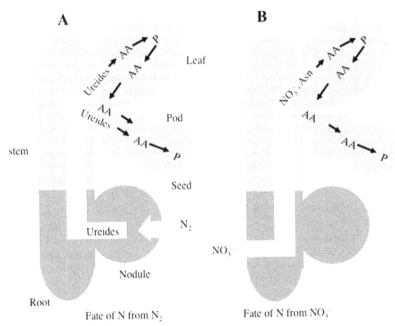

A: Flow of N from N_2 fixed in root nodules.
B: Flow of N from NO_3^- absorbed from roots.
AA: amino acids, P: protein.

Fig. 7. A model of N flow in soybean plant.

1.3 Nitrate inhibition of nodule growth and nitrogen fixation

The inhibitory effects of externally supplied N especially NO_3^- have been reviewed (Streeter, 1988, Harper, 1987). The nitrate inhibition is complex and it cannot be explained by a single mechanism. It has been suggested that there are multiple effects of nitrate inhibition, such as the decrease in nodule number, nodule mass, and N_2 fixation activity, as well as the acceleration of nodule senescence or disintegration (Streeter, 1988, Harter, 1987). In addition, nitrate inhibition of nodules is complex, because the effects of nitrate on nodule formation and growth are influenced by nitrate concentration, placement and treatment period as well as legume species (Harper & Gibson, 1984, Gibson & Harper, 1985, Davidson & Robson, 1986).

Nitrate inhibition is primarily host plant dependent and it is independent of nitrate metabolism of rhizobia (Gibson & Harper, 1984, Carrol & Mathews, 1990). Many hypothesis are proposed for the cause of nitrate inhibition of nodulation and N_2 fixation, i.e. carbohydrate deprivation in nodules (Streeter, 1988, Vessy & Waterer, 1992), feedback inhibition by a product of nitrate metabolism such as glutamine (Neo & Layzell, 1997), asparagine (Bacanambo & Harper, 1996, 1997), and decreased O_2 diffusion into nodules which restricts the respiration of bacteroids (Schuller et al., 1988, Vessey et al., 1988, Gordon et al., 2002). Kanayama and Yamamoto proposed that NO formed from NO_3^- binds to Lb to make nitrosylleghemoglobin and defect the O_2 binding activity (Kanayama & Yamamoto, 1990). On the other hand, Giannakis et al. (1988) suggested that nitrate metabolism does not occur in symbiotic region of soybean nodule, even when a dissimilatory NR is expressed, because of restricted access of nitrate.

Leghemoglobin (Lb) plays a crucial role in N_2 fixation of leguminous nodules by facilitating O_2 supply to the bacteroids. There are four major components of Lb in soybean nodules, Lba, Lbc1, Lbc2, and Lbc3, and different roles are suggested among components (Fuchsman et al., 1976), because Lba has higher affinity for O_2 than has Lbc. The concentrations of Lba and Lbc were separated by Native PAGE (Nishiwaki and Ohyama, 1995). All the four components Lba, Lbc1, Lbc2, and Lbc3 were separately determined by capillary electrophoresis (Sato et al., 1997). The concentration and component ratios in the hypernodulation mutant NOD1-3, NOD2-4, and NOD3-7 from Williams parent, and in En6500 from Enrei parent were compared in relation to their nodulation characteristics. Three mutants (NOD1-3, NOD3-7 and En6500) were controlled by a single recessive allele rj_7, but NOD2-4 was non-allelic mutant to them (Vuong et al. 1996). Plants were hydroponically cultivated in N free solution, and the nodules were separated by size. Concentration and composition of Lb components in the same size nodules were analyzed by gel-electrophoresis and capillary electrophoresis. In all NOD mutants Lb concentration was about 70% of that in the parent Williams, irrespective of nodule size and growth stages. In the hypernodulation mutant En6500, the total Lb concentration was only 25% of that in the parent Enrei, irrespective of nodule size. In Enrei, relative compositions of Lba, Lbc1, Lbc2 and Lbc3 were 36, 26, 18 and 17%, respectively, and very stable irrespective of nodule size. En6500 had relatively equal amounts of each component in which the relative compositions of Lba, Lbc1, Lbc2 and Lbc3 were 30, 22, 22 and 26%. The concentration of Lbc forms in nodules was decreased by addition of nitrate to Enrei plants, but not to En6500. When the nodule morphology was compared among hypernodulation mutant lines and parent lines, we noticed that mutant line had thick cortical regions relative to the comparable parent nodules. The relative volume of symbiotic regions was about 50-60% of total nodule volume in Williams, but it accounted for only 40-50% in NOD mutants.

Sato et al. (2001) investigated the changes in four leghemoglobin components in nodules of NOD1-3 and its parent in the early nodule developmental stage. The hydroponically grown NOD1-3 and Williams were periodically sampled. All the visible nodules were collected from the roots and then the four Lb components in the largest nodules were analyzed by capillary electrophoresis. In NOD1-3 nodule development was faster than those of Williams. Acetylene reduction activity was detected at 19 days after planting in NOD1-3 and at 22 days after planting in Williams. In addition the Lbs were initially detected at 19 days after planting in NOD1-3, a few days earlier than in Williams at 22 days after planting. The Lbcs (Lbc1, Lbc2 and Lbc3) were the main components at the earliest nodule growth stage, and the relative proportion of Lba increased with nodule growth in both NOD 1-3 and Williams. The hypernodulation soybean mutant lines (NOD1-3, NOD2-4, NOD3-7) and the parent Williams and mutant line En6500 and the parent Enrei were cultivated in a sandy dune field in Niigata, and the nodules and root bleeding xylem sap were analyzed at 50, 70, 90 and 120 days after planting (Sato et al., 1998). The number of nodules of the hypernodulation mutant lines was about two to three times higher than that of the parent lines irrespective of sampling date. The concentration of Lb components was measured by capillary electrophoresis. The concentration of Lb components in the hypernodulation mutant lines tended to be lower than in the parents, but the component ratios were not different between mutants and the parents.

It is well recognized that plant growth is affected by various environmental factors, such as temperature, moisture, photoperiod, light intensity and quality, as well as physical, chemical, and biological properties of soil. The degree of nitrate inhibition was affected by soil medium composition with vermiculite and perlite, where the proportion of solid, liquid and gas space was changed (Nishiwaki et al., 1995).

It has been reported in alfalfa that the inhibition of nodulation by nitrate was reduced by medication of ethylene production inhibitor aminoethoxyvinilglycine (Ligero et al., 1991). While the exogenous ethylene inhibited nodulation on the primary and lateral roots of pea (Lee & LaRue, 1992ab). Ethylene is one of the important phytohormone regulating plant growth. Ethylene is produced through oxidative decomposition of 1-aminocyclopropane-1-carbosylic acid (ACC), and silver thiosulfate (STS) is a potent inhibitor of ethylene action in plants (Veen, 1983). Sato et al. (1999c) investigated the effect of ethylene action on soybean nodulation using ACC and STS in relation to the inhibitory mechanism of nitrate using hypernodulation mutant NOD1-3 and the parent Williams. The hypernodulation mutant of soybean NOD1-3 and its parent Williams were cultivated in culture solution with or without NO_3^-, and ACC or STS were added in the solution. The nodule dry weight was decreased by both ACC and STS treatments, however, the ratio in nodule dry weight in total plant dry weight were not significantly influenced by these treatments with or without NO_3^-. Therefore, it was concluded that the decrease in nodule dry weight by ACC or STS was caused by inferior growth. In soybean the depression of nodulation and N_2 fixation by nitrate is not mediated through ethylene action. Schmidt et al. (1999) also reported the independence of ethylene signaling on the regulation of soybean nodulation. Moreover, the nodulation of hypernodulation mutant was not specifically influenced by ACC treatments. This suggests that autoregulation of nodulation may not be involved in ethylene action or transduction pathways in soybean plants.

Recently, defective long-distance auxin transport regulation was reported in the Medicago truncatula super numeric nodules mutant (Van Noorden et al., 2006). However, similar trend is not observed in hypernodulation mutants of soybean. Terakado et al (2005) reported that systemic effect of brassinosteroid on nodule formation in soybean after the foliar application of brassinolide and brassinazaole, the inhibitor of brasinosteroid formation. In addition, they reported that shoot applied polyamines suppressed nodule formation in soybean (Terakado et al., 2006). Suzuki et al. reported that nodule number is controlled by the abscisic acid in Trifolium repense (white clover) and Lotus japonicus (Suzuki et al., 2004).

2. Local effect of nitrate on nodule growth and nitrogen fixation

2.1 Rapid and reversible inhibition of nodule growth and nitrogen fixation by nitrate

Short-term local effect of nitrate supply on nodule formation and nitrogen fixation was evaluated using hydroponically grown soybean plants (cultivar Williams), which were inoculated with *Bradyrhizobium japonicum*, (strain USDA110) (Fujikake et al. 2002. 2003). In the first experiment (Fujikake et al. 2002), the diameter of nodules on the upper part of nodulated soybean roots in a glass bottle was measured with a slide caliper. Nodulated soybean (cv. Williams) plants were hydroponically cultured, and various combinations of one-week culture solution with 5 mM or 0 mM nitrate were applied using 13 days old soybean seedlings during three successive weeks. The treatments were designated as 0-0-0, 5-5-5, 5-5-0, 5-0-0, 5-0-5, 0-5-5 and 0-0-5, where the three sequential numbers denote the nitrate concentration (mM) applied in the first-second-third weeks. The size of the marked individual nodules was measured periodically using a slide caliper. All the plants were harvested after measurement of the acetylene reduction activity (ARA) at the end of the treatments. In the 0-0-0 treatment, the nodules grew continuously during the treatment period. As shown in Fig. 8., individual nodule growth was immediately suppressed after 5 mM nitrate supply. However, the nodule growth rapidly recovered by changing the 5 mM nitrate solution to a 0 mM nitrate solution in the 5-0-0 and 5-5-0 treatments. In the 5-0-5 treatment, nodule growth was completely inhibited in the first and the third weeks with 5mM nitrate, but the nodule growth was enhanced in the second week with 0 mM nitrate. The nodule growth response to 5 mM nitrate was similar between small and large size nodules.

In this experiment nodule numbers are not significantly affected by nitrate treatments (Fig. 9. A), although the nodule weight was significantly affected by the period of nitrate supply (Fig. 9. B), where 5-5-5 and 0-5-5 treatments depressed nodule dry weight about 1/3 of 0-0-0 plants. After the 5-5-5, 5-0-5, 0-0-5 and 0-5-5 treatments, where the plants were cultured with 5 mM nitrate in the last third week, the acetylene reduction activity (ARA) per a plant and ARA per g nodule dry weight (DW) were significantly lower compared with the 0-0-0 treatment (Fig. 10. A,B). On the other hand, the ARA after the 5-0-0 and 5-5-0 treatments was relatively higher than that after the 0-0-0 treatment, possibly due to the higher photosynthate supply associated with the vigorous vegetative growth of the plants supplemented with nitrate nitrogen. It is concluded that both soybean nodule growth and N_2 fixation activity sensitively responded to the external nitrate level, and that these parameters were reversibly regulated by the current status of nitrate in the culture solution, possibly through sensing of the concentration of nitrate or its assimilates in roots and/or nodules.

The nitrate concentration was analyzed in each organ of soybean harvested at the end of the treatment on 34 days after planting. In the plants supplied with 5 mM nitrate during the last week in both the first and second series of treatments (5-5-5, 5-0-5, 0-5-5 and 0-0-5 treatments), the nitrate concentration was significantly high in each organ. Especially the roots and stems accumulated about 9-14 gN kg^{-1}DW and about 5-9 gN kg^{-1}DW nitrate, respectively. On the other hand, the nitrate concentration in roots (0.19 gN kg^{-1}DW), stems (0.03 gN kg^{-1}DW) and nodules (0.11 gN kg^{-1}DW) was fairly low in the 5-5-0 treatment where nitrate was not supplied during the last third week. All the accumulated nitrate during the first and second weeks was reduced and assimilated during the third week of the 0 mM nitrate treatment under the experimental conditions. The nitrate concentration in the nodules was relatively lower than that in the roots and stems, but in the 5-5-5, 0-5-5, 0-0-5 treatments, the nodules accumulated more than 1 gN kg^{-1}DW nitrate.

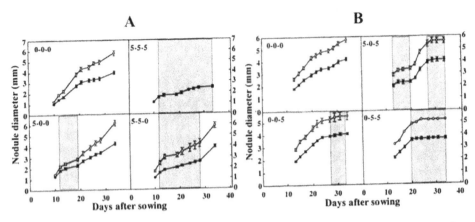

Fig. 8. Changes in nodule diameter of soybean plants with various nitrate treatments. Gray background shows the duration of 5 mM nitrate treatment, and white background shows the 0 mM nitrate. Open circle: large nodules, Closed circle: small nodules.

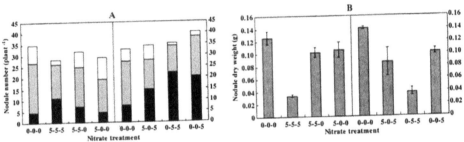

Fig. 9. Number (A) and dry weight (B) of nodules at the end on 34 days after planting. (A) nodule size were indicated by black column (3mm<), gray column (3-5 mm) or white column (<5mm).

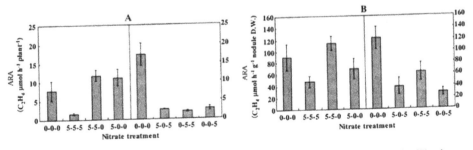

Fig. 10. Acetylene reduction activiety per a plant (A) and per nodule dry weight (B) of nitrate treatments on 34 days after planting.

In the second experiment (Fujikake et al., 2003) and the diameter of individual nodules was measured from 10-24 days after planting using a computer microscope under controlled environmental conditions (Fig. 11.). Photoes (Fig. 11A) and the diameter changes (Fig. 11B) of nodules were shown. A, nodule growth was rapid only under 0 mM nitrate conditions. The diameter of a nodule attached to the primary roots increased from 1 mm to 6 mm for 2 weeks under N free conditions (Fig. 11. Aa, Ba). The increase in nodule diameter was almost completely stopped after 1 d of supplying 5 mM NO_3^- (Fig. 11. Ab, Bb). However, nodule growth quickly returned to the normal growth rate following withdrawal of NO_3^- from the solution (Fig. 11. Ac, Bc).

The morphology of typical nodule slices of soybean observed by an optical microscope is shown in Fig. 12. (A) and the average size of infected cells, uninfected cells and inner cortex cells were measured (Fig. 12. B). It is conspicuous that the infected cells of nodules became larger under 0 mM nitrate condition (a) from 10 to 18 days after planting. On the other hand, the size of infected and uninfected cells and inner cortex cells remained small with 5 mM nitrate solution. Cell growth recovered rapidly after 2 days of 0 mM nitrate (18 days after planting) after 2 days of 5 mM nitrate (16 days after planting) (c). This result indicates that nodule growth at this stage depends on the cell expansion, rather than cell proliferation. The rapid and reversible nodule growth inhibition is caused by nodule cell growth.

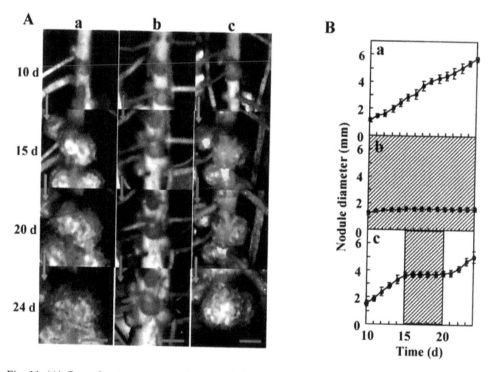

Fig. 11. (A) Growth response of soybean nodules to 0 mM nitrate (blue arrows) or 5 mM nitrate (red arrows) application in the culture solution. (B) Changes in nodule diameter with 0 mM (white background) or 5 mM (hatched background).

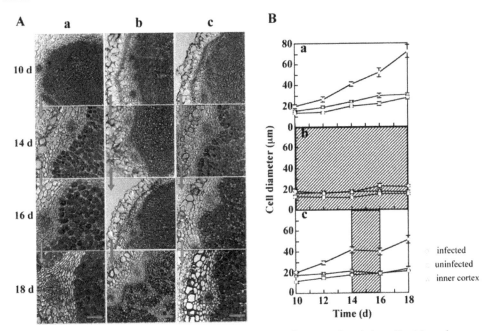

Fig. 12. Effect of 0 mM and 5 mM nitrate treatment on the size of nodule cells. (a) soybean plants were cultured in 0 mM nitrate for 10-18 days. (b) soybean plants were sultured with 5 mM nitrate for 10-18 days. (c) soybean plants were treated with 0 mM from 10-14 days and 5 mM nitrate from 14-16 days and 0 mM nitrate from 16 to 18 days after planting. The size of infected cells, uninfected cells and inner cortex cells were measured in B.

The effect of dark treatment on nodule growth was examined in combination with nitrate treatments for two days followed by normal light/dark conditions (Fig. 13.A.). Under continuous dark conditions, nodule growth maintained on the first day, but depressed on the second day. The reversible depression of nodule growth by NO_3^- was similar to the restriction of the photoassimilate supply under continuous dark conditions for 2 days. The nodule growth with 5 mM nitrate under continuous dark conditions depressed most severely among treatments. When plants were returned to the normal conditions (14 h light and 10h dark) with 0 mM nitrate, all the nodules recovered the growth rate.

The inhibitory effect of 5 mM nitrate was partially alleviated by the addition of 3% sucrose to the culture solution (Fig. 13.B.), suggesting that soybean root nodules are under carbon deficiency.

The positron emitting radioisotope $^{11}CO_2$ was supplied to the first trifoliolate leaves of 29 days after planting for 10 min, then the movement of ^{11}C was monitored by positron-emitting tracer imaging system (PETIS) (Fujikake et al., 2003). Split-root system was made by cutting the primary root of soybean seedling at 24 days after planting. Each split roots was supplied with solution containing with 0 mM or 5 mM NO_3^- for 3 days from 27-29 days after planting. Both sides of split roots were supplied with 0 mM, 0 mM (a), 5mM, 5 mM (b) or 0 mM, 5 mM (c). In the plants with 0 mM, 0 mM or 5 mM 5mM, the ^{11}C assimilated in the first trifoliate was translocated both upward to the young developing apical leaf bud and downward to the whole root system (Fig. 14. A). Very little ^{11}C was transported to the fully developed trifoliates and primary leaves. This means that fully developed leaves are not sink of photoassimilate from other leaves. Compared with the split root system supplied with 0 mM NO_3^- on one side and 5 mM NO_3^- on the other side, the transport rate of ^{11}C was faster in the split-roots supplied with 5 mM NO_3^- than those in 0 mM NO_3^- (Fig. 14. B, C). This result indicates that when NO_3^- is supplied to a part of the roots, photoassimilate flow become faster in this part.

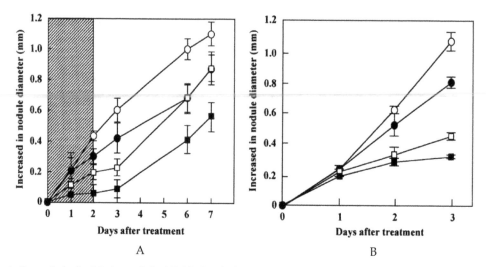

A: Open circle: 0 mM nitrate, light 14h/dark 10h. filled circle: 0 mM nitrate, continuous dark for 2 days. Open square: 5 mM nitrate, light 14h/dark 10h. filled square: 5 mM nitrate, continuous dark for 2 days. B: Effect of 3% sucrose on nodule growth (B). Open Circle: 3% sucrose + 0 mM nitrate. Filled circle: 0% sucrose + 0 mM nitrate. Open square: 3% sucrose + 5 mM nitrate. filled square: 0% sucrose + 5 mM nitrate.

Fig. 13. Nodule growth of soybean plants grown with 0 mM or 5 mM nitrate under light/dark conditions

Quantitative evaluation was conducted using ^{14}C as a tracer for the plants supplied either with 0 mM or 5 mM NO_3^- for one day before supplying $^{14}CO_2$. Whole shoot of a plant at 22 days after planting was exposed to the $^{14}CO_2$ for 120 min using circulation system. After $^{14}CO_2$ feeding, plant samples were immediately dried and the radioactivity in leaves, stems, nodules and roots were determined using Liquid Scintillation Counter. By supplying 5 mM NO_3^-, the partitioning to the underground part were almost the same in 0 and 5 mM nitrate treatments, but the ^{14}C partitioning to nodule decreased from 9.1 % to 4.3%, while that to the roots increased from 5.2 % to 9.1% (Fig. 15).

These results indicate that the decrease in photoassimilate supply to nodules may be involved in the quick and reversible nitrate inhibition of soybean nodule growth and N_2 fixation activity (Fig. 16.). The decrease in starch concentration in nodules (Vessey et al., 1988, Gordon et al., 2002) and the down-regulation of sucrose synthase transcript within 1 day of nitrate treatment (Gordon et al., 2002) may imply that NO_3^- reduces photosynthesis flow into nodules and sucrose utilization in nodules.

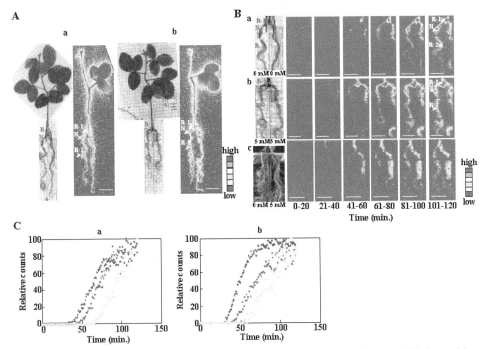

Fig. 14. ¹¹C translocation to the split root systems from first trifoliate leaves of 29 days old plants. (A) Images of the distribution of ¹¹C in soybean by Bioimaging analyzer. (B) The time course for the accumulation of radioactivity as shown by PETIS. (C) The accumulation of radioactivity for the point of R-1 (blue), N (red) and R-2 (yellow).

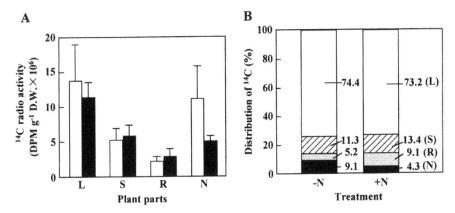

Fig. 15. Partitioning of ¹⁴C labeled photoassimilate in soybean plants. (A) Radioactivity per g dry weight of each part after one day of 2 h ¹⁴CO₂ feedings to a whole shoot, with 0 mM nitrate (white column) and 5 mM nitrate (black column). (B) Distribution of ¹⁴C among organs with 0 mM (-N) or 5 mM (+N) treatment for one day. (L): leaves, (S): stems, (R): roots, (N): nodules.

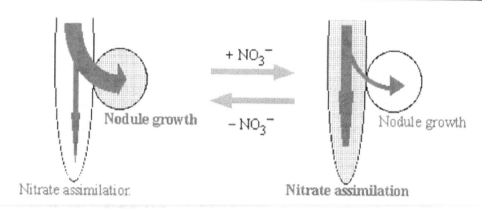

Fig. 16. A model of photoassimilate transport to nodule and roots under no nitrate or presence of nitrate.

2.2 The routes of nitrate entry into nodules

It is well known that nitrate inhibit nodule growth and nitrogen fixation activity, but the routes of nitrate entry into nodule has not fully been understood. It is postulated that there are several routes in nodules. First, NO_3^- is absorbed from the roots and transported to nodules through the xylem. However, xylem transport of NO_3^- to nodules is negligible, because nitrate accumulation is very low in separated nodules in the upper part of roots, when NO_3^- was supplied in the lower roots. In addition the role of xylem in nodule is the export of assimilated nitrogen from nodules to shoots, rather than the import of water and minerals from roots. Most of water and minerals are supplied from shoots to nodules though phloem. Second route is NO_3^- supplied via phloem. However, usually NO_3^- concentration in phloem is very low due to nitrate reduction in leaves. Third is NO_3^- is taken up from the subtending roots and transported from root cortex to the nodule cortex via symplastic pathway (Streeter 1993). Forth, NO_3^- is absorbed from nodule surface.

The nitrate transport pathway into soybean nodules were investigated using tungstate (WO_4^{2-}) and $^{15}NO_3^-$ as a tracer (Mizukoshi et al. 1995). Tungstate was used as an anion tracer as an analogue of nitrate (NO_3^-). The distribution of tungsten (W) was observed by an Electron Probe X-ray Microanalysis (EPMA). At 3 days after 1 mM tungstate treatment, a large amount of W accumulated in the root cortex, and the import of W into stele was restricted (Fig. 17A, C). It is well known that there is a barrier in endodermis between cortex and stele, where water movement is not allowed. Therefore, solutes should pass through endodermis through inside the cells via symplastic pathway. In addition, the movement of W inside the nodule was negligible (Fig. 17 C). This result suggests that the external anions cannot be readily enter into the cortex of nodules through appoplastic pathway by diffusion. In contrast, nodulated roots were treated with 1.7 mM $^{15}NO_3^-$ for one day, a relatively large amount of nitrate was accumulated in nodule cortex, although NO_3^- and ^{15}N were negligible in the symbiotic region. This result indicates that nitrate can be absorbed from the nodule surface into cytoplasm of nodule epidermis, and it is transported by symplastic pathway through plasmodesmata, and accumulated in the nodule cortex cells.

As shown in Fig. 19., the accumulation of NO_3^- in nodule cortex may be involved in the restriction of O_2 permeability into symbiotic resion, then it decreases the nitrogen fixation and nodule growth. Witty and Minchin (1990) reprted that the NO$_3^-$ treatment decreased the air space within a oxygen barrier in inner cortex of nodules.

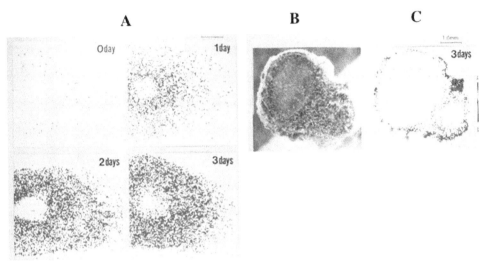

Fig. 17. Distribution of tungsten (W) in soybean root at 0, 1, 2, 3 days after 1 mM tungstate treatment (A). Distribution of tungsten (W) in soybean nodule with root at 3 days after 1 mM tungstate treatment (A). The W concentration is higher indicated by red, yellow , green, blue, and white, in this sequence.

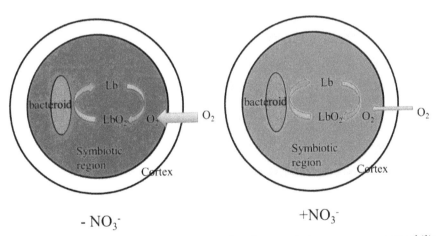

Fig. 18. Effect of nitrate accumulation in cortex of soybean nodule on oxygen permeability and nitrogen fixation activity.

Sato et al. (1999) analyzed nitrogen absorption and translocation in non-nodulated and nodulated soybean plants cultivated with 0 mM or 1 mM nitrate (Fig. 19.). The radioactivity was measured in a first trifoliate leaf after addition of $^{13}NO_3^-$ supply to the root solution. The relative radioacrivity from 0 to 40 min after $^{13}NO_3^-$ supply were similar between non-nodulated and nodulated soybean with 0 or 1 mM nitrate (Fig. 20.). This suggests that nodulation does not change the absorption and transport pattern of nitrate absorbed in the roots. However, quantitative measurement using stable isotope $^{15}NO_3^-$, total amount of ^{15}N was higher in non-nodulated soybean than nodulated soybean, especially with 1 mM $^{15}NO_3^-$, this is due to the increase in the root mass in these plants.

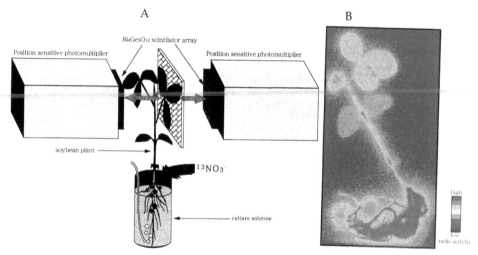

A: Positron Emitting Tracer Imaging System. B: Image of ^{13}N radioisotope in non-nodulated soybean plant (T202).

Fig. 19. Real time observation of radioactivity in the first trifoliate of soybean plant after $^{13}NO_3^-$ was supplied to the root solution.

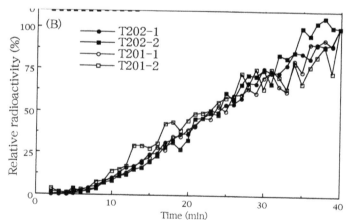

Fig. 20. Changes in relative radio activity in leaf of nodulated (T-202) and non-nodulated (T-201) isolines of soybean after $^{13}NO_3^-$ was supplied to the root solution. The radioactivity at 40 min is normalized as 100%.

3. Systemic and long-term effect of nitrate on nodule growth and nitrogen fixation

3.1 The effect of 0mM or 5 mM nitrate application in upper and lower part of roots

Local and systemic effects by nitrate on nodulation have been reported in leguminous plants. The local effect of nitrate inhibition was shown in split-root experiments where root systems had been separated into two equivalent parts. The strong and rapid nitrate inhibition of nodule growth and N_2 fixation activity is restricted in the nodules attached to the root portions that are in direct contact with nitrate; and no or milder inhibition is induced in the other part of the root system receiving no nitrate (Tanaka et al., 1985). However, some systemic inhibition of nitrate on nodulation and nitrogen fixation has also been observed with a high concentration of nitrate in clover (Silsbury et al., 1986).

We investigated the local and systemic effects of continuous supply of NO_3^- by using horizontal sprit-root system in two-layered pot system, where the lower part of roots were supplied with culture solution containing 1mM NO_3^- in the lower pot, and the upper roots were in the vermiculite medium with N-free culture solution in the upper pot (Ohyama et al., 1993). The soybean plants (cv. Williams and Norin No.2) were cultivated with 0 mM or 1 mM NO_3^- solution in the lower pot, and harvested at maturing (R7) stage (Fher et al., 1971). In this stage, there are no nodules remained in the lower part of roots. The dry weight of shoot and upper part of roots were almost the same between 0 mM and 1 mM NO_3^- supply, but nitrate treatment decreased the dry weight of nodules attached in the upper part of roots in both varieties. This result indicates that continuous long-term supply of NO_3^- may impose systemic inhibition of nodulation in soybean plants.

Systemic and local effects of long-term application of nitrate on nodule growth and N_2 fixation in soybean plants were more precisely investigated using two layered pot system. Four treatments were imposed i.e., 0/0, 0/5, 5/0 and 5/5, with the 0 mM or 5 mM NO_3^- treatment in upper pot/ lower pot, respectively. The plants were harvested at the initial flowering (R1) stage and pod setting (R4) stage, and the effect of nitrate placement on nodule number, nodule growth, and N_2 fixation activity in the upper and lower pots were elucidated (Yashima et al., 2003).

The development of the root system in the lower pots was quite different between 0 and 5 mM NO_3^- in the lower pot (Fig. 21.). The root length was longer in 0 mM treatment in lower pot (0/0, 5/0), but a bunch of short lateral roots was formed in the solution with 5 mM NO_3^- in lower pot (0/5, 5/5). In the lower pot where the nodules were in direct contact with 5 mM NO_3^-, the inhibition on the nodule number, nodule size and N_2 fixation was conspicuous. Systemic and local effect on nodule number per a plant did not occur in the upper nodules in vermiculite. On the other hand, systemic inhibition on the nodule dry weight and N_2 fixation activity in the upper pot was apparent. The 5/5 treatment depressed the nodule growth and nitrogen fixation activity in the upper nodules. Nitrate accumulation was observed only in the part of roots and nodules in direct contact with 5 mM NO_3^- either in the upper or lower pot. The concentration of total amino acids was higher in the lower roots in 0/5 treatment than those in 0/0 treatment, however, that was almost the same level in the roots and nodules of the upper pot both at R1 and R4 stage. The soluble sugar concentration in the lower roots in 0/5 treatment was lower than that in the 0/0 treatment. The similar trend was observed in the upper roots of 0/5 treatment, suggesting that the absorption of NO_3^- from the lower roots decrease sugar concentration in both lower roots in direct contact with nitrate, and the upper roots not contact with NO_3^-.

A   B   C   D

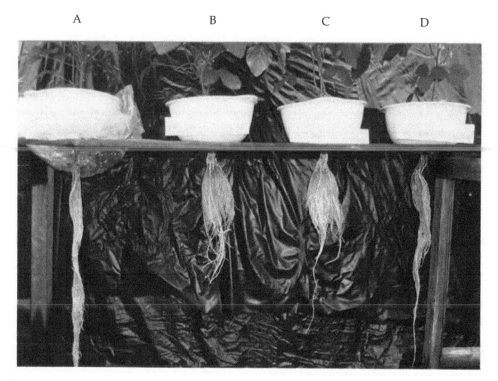

Fig. 21. Root system in the lower pot treated with 0/0 (A), 0/5 (B), 5/5 (C) and 5/0 (D) treatments (upper pot mM nitrate / lower pot mM nitrate).

3.2 Long-term effects of 0 mM, 1 mM, 5 mM nitrate application in lower part of roots

Long-term effect of NO_3^- application from the lower part of roots on the nodulation of the upper part of roots was further investigated in relation to concentration and treatment period (Yashima et al. 2005). The solution with 0 mM, 1mM or 5 mM NO_3^- was supplied from transplanting to two-layered pot system at 14 days after planting to R7 stage. Five treatments were imposed that 0-0 treatment (continuous 0 mM NO_3^-), 1-1 treatment (continuous 1 mM NO_3^-), 5-5 treament (continuous 5 mM NO_3^-), 0-5 treatment (0 mM until R3 then 5 mM NO_3^-), and 5-0 treatment (5 mM until R3 then 0 mM NO_3^-).

Total plant dry weight and seed dry weight at R7 stage was the highest in 5-5 treatment, intermediate in the 1-1, 5-0, 0-5 treatments, and lowest in the 0-0 treatment (Fig. 22.).

Fig. 23. shows the nodule number per a plant classified with nodule diameter. Nodule number in the upper pot was higher in 5-5 and 1-1 treatments than 0-0 treatment, although proportion of the small nodules under 4 mm diameter was higher in 5-5 treatment. The nitrate supply in the lower pot increased the total nodule number in the upper roots, although decreased the number of nodules in the lower roots.

The value of the nodule dry weight per a plant (Fig.24.) and N_2 fixation activity (acetylene reduction activity) per a plant (Fig. 25) and that per nodule dry weight (Fig. 26.) were lowest in the 5-5 treatment. Interestingly, the nodule dry weight in the upper roots was highest in the plants with 1-1 treatment, which exceeded the 0-0 treatment (Fig. 24.). The acetylene reduction activity per a plant of the upper nodules at R3 stage was also the highest in the 1-1 treatment (Fig. 25.). This was due to the nodule dry weight and not ARA per dry weight of nodules. These results indicated that continuous supply of low concentration of NO_3^- from the lower roots does not inhibit the nodule growth and N_2 fixation activity, but it can promote nodulation and N_2 fixation. Fig. 27. shows an example of soybean root systems cultivated with continuous supply of 1 mM nitrate at R3 stage. Nodulation was enhanced in the upper roots by supplying 1mM NO_3^- from the lower roots (Fig. 27. Ba), where the nodulation was severely depressed in the lower pot (Fig. 27. Bb).

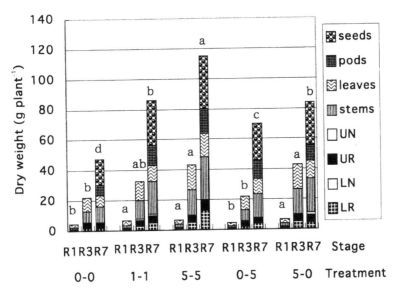

Different alphabet on the column means the statistical difference (P<0.05) between treatments.

Fig. 22. Changes in dry weight of each part of soybean plants at R1 (initial flowering stage; 40 days after planting), R3 (initial pod setting stage, 60 days after planting), R7 (maturing stage; 100 days after planting) with various nitrate treatments. 0-0, 1-1, 5-5, 0-5, 5-0 indicates mM nitrate concentration from transplanting-R3 stage, and R3-R7 stage respectively.

Withdrawal of 5 mM NO_3^- after R3 stage in 5-0 treatment markedly enhanced nodule growth and acetylene reduction activity at R7 stage when the values of both parameters decreased in the other treatments. The nitrate concentration of the nodules attached to the upper roots was very low, including continuous supply of high concentration of NO_3^- in 5-5 treatment. This result indicates that the inhibitory effect of 5 mM NO_3^-, or promotive effect of 1 mM NO_3^- was not directly controlled by nitrate itself, but was mediated through some systemic regulation.

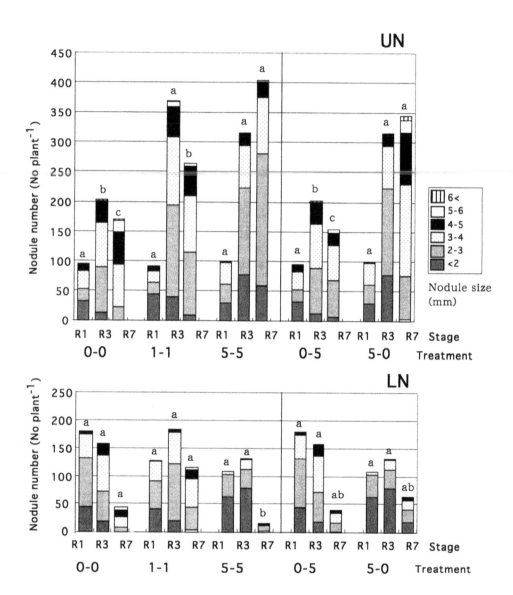

UN: upper nodules, LN: lower nodules.
Different alphabet on the column means the statistical difference (P<0.05) between treatments.

Fig. 23. Nodule number per a plant with various nitrate treatment in the lower pot. Size distribution is shown.

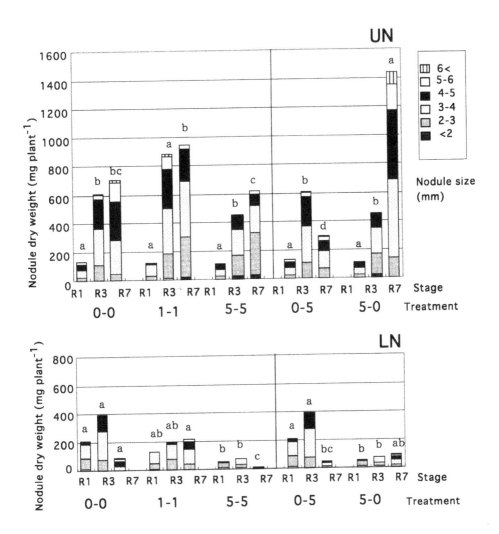

UN: upper nodules, LN: lower nodules.
Different alphabet on the column means the statistical difference (P<0.05) between treatments.

Fig. 24. Nodule dry weight per a plant with various treatment of nitrate in a lower pot.

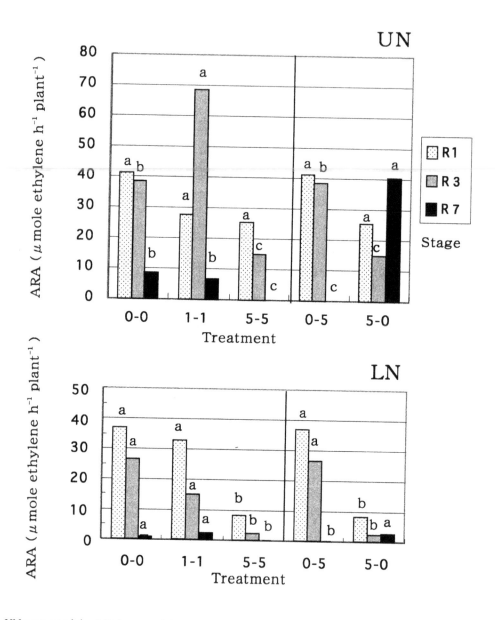

UN: upper nodules, LN: lower nodules.

Different alphabet on the column means the statistical difference (P<0.05) between treatments.

Fig. 25. Nitrogen fixation activity per a plant in soybean plants treated with various nitrate supply from lower pot.

b)

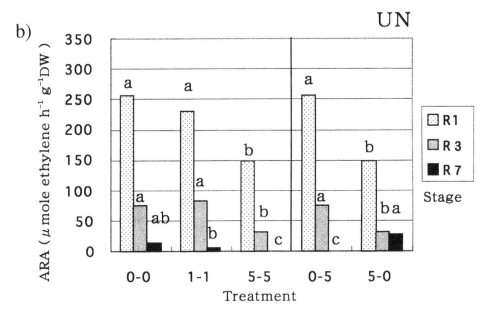

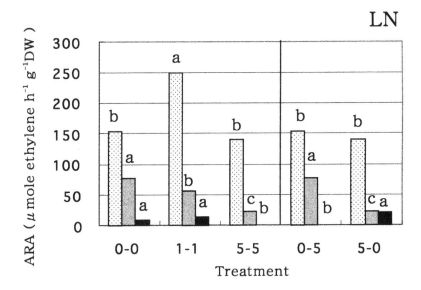

Different alphabet on the column means the statistical difference (P<0.05) between treatments.

Fig. 26. Nitrogen fixation activity per gram nodule dry weight in soybean plants treated with various nitrate supply from lower pot. UN: upper nodules, LN: lower nodules.

A   B

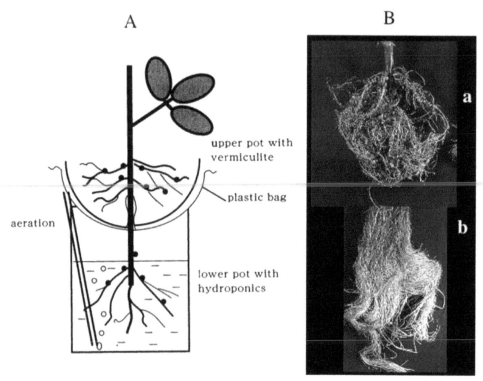

upper pot with
vermiculite

plastic bag

aeration

lower pot with
hydroponics

A: Apparatus for two-layered pot.
B: photograph of root system in the upper roots with 0 mM nitrate (a) and the lower roots with 1 mM nitrate(b).

Fig. 27. A vertical split root system used in long-term effect of nitrate from lower pot.

4. Conclusion

In this chapter, we focus on the effects of nitrate supply on nodule growth and nitrogen fixation in soybean plants. First, we found the rapid and reversible inhibition of 5mM nitrate on nodule growth and nitrogen fixation activity in soybean plants. When young soybean plant grown in hydroponic culture was supplied with 5 mM nitrate solution, the nodule growth was completely stopped after one day of application. The culture solution was changed back to nitrogen free solution the nodule growth and nitrogen fixation activity were recovered at one day after changing. The inhibitory effect by nitrate may be due to the changes in photoassimiate supply from nodule to the roots by the experiments exposing the shoot to positron emitting radioisotope ^{11}C or radioisotope ^{14}C labeled CO_2.

Second, the effect of long-term application of nitrate was evaluated by the vertical split root experiments using two-layered pot, which separates the upper and lower parts of the root system. Both direct and systemic inhibitions were observed for nodule dry weight and nitrogen fixation activity with long-term supply of 5 mM nitrate. Severe inhibition in the root part in direct contact with 5mM nitrate solution was observed, although milder depression was in the separate parts not in direct contact with nitrate. Nitrate accumulation in nodule was observed only in the root part in direct contact with 5 mM nitrate. When the 1mM concentration of nitrate was continuously supplied from the lower roots, the nodule growth and nitrogen fixation activity in the upper roots were promoted compared with control plants supplied with nitrogen free solution. When 5 mM nitrate supply has stopped from pod setting stage, nodule weight and nitrogen fixation activity were recovered and exceeded over control plant at maturing stage.

The routes of nitrate entry into nodules was investigated. It was suggested that most of nitrate is absorbed through the surface of the nodule, but little is transported from the lower part of roots or shoots, either via xylem or phloem. The accumulation of nitrate in the cortex of nodule may inhibit respiration by decreasing O_2 permeability or suppress the photoassimilate import to the central symbiotic region of nodules, where bacteroid reside and fix nitrogen.

From the characteristics of nitrate effects on soybean nodulation and nitrogen fixation, we have developed a new fertilizer method "deep placement of slow release nitrogen fertilizer" for promoting soybean seed yield without inhibiting nodulation and nitrogen fixation (Takahashi et al., 1991, 1992, 1999, Tewari et al., 2010).

5. References

Atkins, C.A. (1991). Ammonia assimilation and export of nitrogen from the legume nodule. In *Biology and Biochemistry of Nitrogen Fixation*, Elsevier Science Publishers, 293-319

Bacanamwo, M. & Harper, J.E. (1996). Regulation of nitrogenase activity in *Bradyrhizobium japonicum*/soybean symbiosis by plant N status as determined by shoot C:N ratio. *Physiol. Plant.*, 98, 529-538.

Bacanamwo, M. & Harper, J.E. (1997). The feedback mechanism of nitrate inhibition of nitrogenase activity in soybean may involve asparagines and/or products of its metabolism. *Physiol. Plant.* 100, 371-377.

Bloom, A.J. (2011). Energetics of Nitrogen Acquisition, *Annual Plant Reveiws*, 42, 63-81

Carroll, B.J. & Mathews, A. (1990). Nitrate inhibition of nodulation in legumes. In: Gresshoff PM, ed. *Molecular Biology of Symbiotic Nitrogen Fixation*. Boca Raton, FL: CRC press, 159-180.

Davidson, I.A. & Robson, M.J. (1986). Effect of contrasting patterns of nitrate application on the nitrate uptake, N_2-fixation, nodulation and growth of white clover. *Ann. Botany* 57, 331-338.

Fehr, W.R., Caviness, C.E., Burmood, D.T., and Pennington, J.S., 1971. Stage development description for soybean *Glycine max* (L.) Merill. *Crop Sci.*, 11, 929-931.

Fleurat-Lessard, P., Michonneau, P., Maeshima, M., Drevon, J-J. & Serraj, R. (2005). The Distribution of Aquaporin Subtypes (PIP1, PIP2 and g-TIP) is Tissue Dependent in Soybean (*Glycine max*) Root Nodules, *Annals of Botany* 96: 457–460, 2005

Fuchsman, W.H., Barton, C.R., Stein, M.M., Thomson, J.T. & Willett, R.M., (1976). Leghemoglobin: Different roles for different components?, *BBRC*, 68, 387-392

Fujikake, H., Yashima, H., Sato, T., Ohtake, N., Sueyoshi, K., and Ohyama, T., (2002) Rapid and reversible nitrate inhibition of nodule growth and N_2 fixation activity in soybean (*Glycine max* (L.) Merr.). *Soil Sci. Plant Nutr.,* 48, 211-217

Fujikake, H., Yamazaki, A., Ohtake, N., Sueyoshi, K., Matsuhashi, S., Ito, T., Mizuniwa, C., Kume, T., Hashimoto, S., Ishioka, N-S., Watanabe, S., Osa, A., Sekine, T., Uchida, H., Tsuji, A. & Ohyama, T. (2003). Quick and reversible inhibition of soybean root nodule growth by nitrate involves a decrease in sucrose supply to nodules. *Journal of Experimental Botany,* 54, 1379-1388

Fujimaki, S., Ishii, S. & Ishioka N-S. (2010). Non-invasive imaging and analyses of transport of nitrogen nutrition in intact plants using a positron-emitting tracer imaging system (PETIS). Nitrogen Assimilation in Plants, Ed. Ohyama T. & Sueyoshi K. Research Signpost, Kerala

Giannakis, C., Nicholas, D.J.D. & Wallace, W. (1988). Utilization of nitrate by bacteroids of *Bradhyrhizobium japonicum* in the soybean root nodule. *Planta,* 174, 51-58

Gibson, A.H. & Harper, J.E. (1985). Nitrate effect on nodulation of soybean by *Bradyrhizobium japonicum. Crop Science,* 25, 497-501.

Gordon, A.J., Skøt, L., James, C.L. & Minchin, F.R., (2002) Short-term metabolic response of soybean root nodules to nitrate. *Journal of Experimental Botany.,* 53, 423-428.

Harper, J.E. (1974). Soil and symbiotic nitrogen requirements for optimum soybean production. *Crop Sci.* 14, 255-260.

Harper, J.E. & Gibson, A.H., (1984). Differential nodulation tolerance to nitrate among legume species, *Crop Sci.* 14, 255-260

Harper, J.E. (1987). *Nitrogen metabolism. Soybeans: Improvement, production and Uses.* 2nd ed. *Agronomy Monograph* no.16. ASA-CSSA-SSSA. 497-533.

Hunt, S. & Layzell, D.B., (1993). Gas exchange of legume nodules and the regulation of nitrogenase acrivity, *Annu. Rev. Plant Physiol. and Plant Mol. Biol.,* 44, 483-511.

Ishii, S., Suzui. N., Ito, S., S-Ishioka, N., Kawachi, N., Ohtake, N., Ohyama, T. & Fujimaki, S. (2009). Real-time imaging of nitrogen fixation in an intact soybean plant with nodules using ^{13}N-labbeled nitrogen gas. *Soil Sci. Plant Nutr.* 55, 660-666

Ito, S., Ohtake, N., Sueyoshi, K. & Ohyama, T. (2006). Allocation of photosynthetic products in soybean during the early stages of nodule formation. *Soil Sci. Plant Nutr.,* 41, 438-443

Kanayama, Y. & Yamamoto, Y. (1990). Inhibition of nitrogen fixation in soybean plants supplied with nitrate I. Nitrite accumulation and formation of nitrosylleghemoglobin in nodules. *Plant Cell Physiol.* 31, 341-346

Lee, K.H. and LaRue, T.A., (1992). Ethylene as a possible mediator of light-and nitrate-indulced inhibition of nodulation of *Pisum sativum* L. cv Sparkle. *Plant Physiol.,* 100, 1334-1338

Ligero, F., Caba, J.M., Lluch, C., Oliveres, J., (1991). Nitrate inhibition of nodulation can be overcome by the ethylene inhibitor aminoethoxyvinylglycine. *Plant Physiology* 97, 1221-1225.

Mizukoshi, K., Nishiwaki, T., Ohtake, N., Minagawa, R., Ikarashi, T., & Ohyama, T. (1995). Nitrate transport pathway into soybean nodules traced by tungstate and $^{15}NO_3^-$. *Soil Sci. Plant Nutr.*, 41, 75-88

Neo, H.H. & Layzell, D.B. (1997). Phloem glutamine and the regulation of O_2 diffusion in legume nodules. *Physiologia. Plantarum* 113, 259-267.

Nishiwaki, T., Mizukoshi, K., Kobayashi, K., Ikarashi, T. & Ohyama, T. (1995). Effect of culture medium compositions on nodulation, nitrogen fixation and growth of soybean plant. *Bull. Facul. Agric. Niigata Univ.*, 47, 73-83

Ohtake, N., Nishiwaki, T., Mizukoshi, K., Minagawa, R., Takahashi, Y., Chinushi, T. & Ohyama ,T. (1995). Amino acid composition in xylem sap of soybean related to the evaluation of N_2 fixation by the relative ureide method. *Soil Sci.Plant Nutr.*, 41, 95-102.

Ohyama, T. & Kumazawa, K. (1978). Incorporation of ^{15}N into various nitrogenous compounds in intact soybean nodules after exposure to $^{15}N_2$ gas. *Soil Sci.Plant Nutr.*, 24, 525-533.

Ohyama, T. & Kumazawa, K. (1979). Assimilation and transport of nitrogenous compounds originated from $^{15}N_2$ fixation and $^{15}NO_3^-$ absorption. *Soil Sci.Plant Nutr.*, 25, 9-19

Ohyama, T. & Kumazawa, K. (1980a). Nitrogen assimilation in soybean nodules I. The role of GS/GOGAT system in the assimilation of ammonia produced by N_2 fixation. *Soil Sci.Plant Nutr.*, 26, 109-115

Ohyama, T. & Kumazawa, K. (1980b). Nitrogen assimilation in soybean nodules II. $^{15}N_2$ assimilation in bacteroid and cytosol fractions of soybean nodules. *Soil Sci.Plant Nutr.*, 26, 205-213

Ohyama, T. & Kumazawa, K. (1980c). Nitrogen assimilation in soybean nodules III. Effect of rhizaosphere pO_2 on the assimilation of $^{15}N_2$ in nodules attatched to intact plants. *Soil Sci.Plant Nutr.*, 26, 321-324

Ohyama, T. Owa N, Fujishima Y & Kumazawa, K. (1981a). Nitrogen assimilation in soybean nodules IV. Allantoin formation and transport in relation to supply with various forms of combined nitrogen. *Soil Sci.Plant Nutr.*, 27, 55-64

Ohyama, T. & Kumazawa, K. (1981b). Nitrogen assimilation in soybean nodules V. Possible pathway of allantoin synthesis in soybean nodules. *Soil Sci.Plant Nutr.*, 27, 111-114

Ohyama, T. (1983). Comparative studies on the distribution of nitrogen in soybean plants supplied with N_2 and NO_3^- at the pod filling stage. *Soil Sci.Plant Nutr.*, 29, 133-145.

Ohyama, T. (1984). Comparative studies on the distribution of nitrogen in soybean plants supplied with N_2 and NO_3^- at the pod filling stage. II. Assimilation and transport of nitrogenous constituents. *Soil Sci.Plant Nutr.*, 30, 219-229.

Ohyama, T., Mizukoshi, K., Nishiwaki, T. (1993a). Distribution of ureide originated from nodules attached to the upper roots and nitrate derived from lower roots in soybean plants cultivated by double piled pots. *Bull. Facul. Agric. Niigata Univ.* 45, 107-116.

Ohyama, T., Ohtake, N., Sueyoshi, K., Tewari, K., Takahashi, Y., Ito, S., Nishiwaki, T., Nagumo, Y., Ishii, S. & Sato, T. (2009). *Nitrogen Fixation and Metabolism in Soybean Plants*, Nova Science Publishers, Inc., New York.

Sato, T., Nishiwaki, T., Ohtake, N. & Ohyama, T. (1997). Determination of concentration of soybean nodule leghemoglobin components by capillary electrophoresis. *Jpn J. Soil Sci. Plant Nutr.*, 68, 521-526

Sato, T., Yashima, H., Ohtake, N., Sueyoshi, K., Akao, S., Harper, J.E., & Ohyama, T. (1998). Determination of leghemoglobin components and xylem sap composition by capillary electrophoresis in hypernodulation soybean mutants cultivated in the field. *Soil Sci. Plant Nutr.*, 44, 635-645

Sato, T., Yashima, H., Ohtake, N., Sueyoshi, K., Akao, S. & Ohyama, T (1999a). Possible involvement of photosynthetic supply in changes of nodule characteristics of hypernodulating soybeans. *Soil Sci. Plant Nutr.*, 45, 187-196

Sato, T., Ohtake, N., Ohyama, T., Ishioka, N.S., Watanabe. S., Osa, A., Sekine, T., Uchida, H., Tsuji, A., Matsuhashi, S., Ito, T. & Kume, T. (1999b). Analysis of nitrate absorption and transport in non-nodulated and nodulated soybean plants with $^{13}NO_3^-$ and $^{15}NO_3^-$. *Radioisotopes*, 48, 450-458.

Sato, T., Nishiwaki, T., Ohtake, N., Sueyoshi, K., & Ohyama, T. (1999c). Non-involvement of ethylene action on the nitrate inhibition of nodulation in hypernodulation soybean mutant and its parent cv. Williams. *Bull. Fac. Agric. Niigata Univ.*, 51, 121-130

Sato, T., Onoma, N., Fujikake, H., Ohtake, N., Sueyoshi, K. & Ohyama, T. (2001). Changes in four leghemoglobin components in nodules of hypernodulating soybean (*Glycine max* [L] Merr.) mutant and its parent in the early nodule developmental stage. *Plant and Soil*, 237, 129-135

Silsbury, J.H., Catchpoole, D.W.& Wallance ,W. (1986). Effects of nitrate and ammonium on nitrogenase (C_2H_2 reduction) activity of swards of subterranean clover, *Trifolium subterraneum* L. *Australian Journal of Plant Physiology* 13, 257-73.

Schimidt, J.S., Harper, J.E., Hoffman, T.K., Bent, A.F. (1999). Regulation of soybean nodulation independent of ethylene signaling. *Plant Physiology* 119, 951-959.

Schubert, K.R., (1986). Products of biological nitrogen fixation in higher plants; symbiosis, transport and metabolism. *Annu. Rev. Plant Physiol.*, 37, 539-574

Schuller, K.A., Minchin, F.R., Gresshoff, P.M. (1988). Nitrogenase activity and oxygen diffusion in nodules of soybean cv. Bragg and a supernodulating mutant: effects of nitrate. *Journal of Experimental Botany* 39, 865-877.

Serraj, R., Fleurat-Lessard, P., Jaillard, B. & Drevon, J.J. (1995). Structural changes in the inner-cortex cells of soybean root-nodules are induced by short-term exposure to high salt or oxygen concentrations. *Plant, Cell and Environment*, 18: 455–462.

Serraj, R., Frangne, N., Maeshima, M., Fleurat-Lessard, P. & Drevon, J.J. (1998). A γ-TIP cross-reacting protein is abundant in the cortex of soybean N_2-fixing nodules. *Planta* 206: 681–684

Streeter, J.G. (1988). Inhibition of legume nodule formation and N_2 fixation by nitrate. *CRC Crit. Rev. Plant Sci.*, 7, 1-23

Streeter, J.G. (1993). Translocation-A key factor limiting the efficiency of nitrogen fixation in legume nodules. *Plant Physiol.*. 87 616-623

Suzuki, A., Akune, M., Kogiso, M., Imagama, Y., Osuki, K.,Uchiumi, T., Higashi, S., Han, S., Yoshida, S., Asami, T. & Abe, M. (2004). Control of nodule number by the phytohormone abscisic acid in the roots of two leguminous species. *Plant Cell Physiol.*, 45, 914-922

Tajima, S., Nomura, M & Kouchi, H. (2004). Ureide biosynthesis in legume nodules. *Frontiers in Bioscience*, 9, 1374-1381

Takahashi, Y., Chinushi, T., Nagumo, Y., Nakano, T., and Ohyama, T., 1991. Effect of deep placement of controlled release nitrogen fertilizer (coated urea) on growth, yield and nitrogen fixation of soybean plants. *Soil Sci. Plant Nutr.*, 37, 223-231.

Takahashi, Y., Chinushi, T., Nakano, T. & Ohyama, T. (1992). Evaluation of N_2 fixation and N absorption activity by relative ureide method in field grown soybean plants with deep placement of coated urea. *Soil Sci. Plant Nutr.*, 38, 699-708.

Takahashi, Y. & Ohyama, T. (1999). Technique for deep placement of coated urea fertilizer in soybean cultivation. *JARQ*, 33, 235-242.

Tanaka, A., Fujita, K. & Terasawa, H. (1985). Growth and dinitrogen fixation of soybean root system affected by partial exposure to nitrate. *Soil Sci. Plant Nutr.* 31, 637-645.

Terakado, J., Fujihara, S., Gott S., Kutatai, R., Suzuki, Y., Yoshida, S. & Yoneyama, T., (2005). Systemic effect of a brassinosteroid on root nodule formation in soybean as revealed by the application of brassinolide and brassinozaole, *Soil Sci. Plant Nutr.*, 51, 389-395.

Terakado, J., Yoneyama, T. & Fujihara, S., (2006). Shoot-applied polyamines suppress nodule formation in soybean (*Glycine max*), *J. Plant Physiol.*, 163, 497-505.

Tewari, K., Nagumo, Y., Takahashi, Y., Sueyoshi, K., Ohtake, N. & Ohyama, T. (2010). A New Tecnology of Deep Placement of Slow Release Nitrogen Fertilizers for Promotion of Soybean Growth and Seed Yield. NOVA Science Publishers

Van Noorden, G.E., Ross, J.J., Reid, J.B., Rolfe, B.G. & Mathesius, U. (2006). Defective long-distance auxin transport regulation in the Medicago truncatula super numeric nodules mutant[w]. *Plant Physiol.*, 140, 1494-1506

Vessey, J.K., Walsh, K.B. & Layzell, D.B. (1988). Can a limitation in phloem supply to nodules account for the inhibitory effect of nitrate on nitrogenase activity in soybean? *Physiol. Plant.*, 74, 137-146

Vessey, J.K. & Waterer, J. (1992). In search of the mechanism of nitrate inhibition of nitrogenase activity in legume nodues: Recent development. *Physiologia Plantarum* 84, 171-176.

Vuong, T.D., Nickell, C.D. & Harper, J.E., (1996). Genetic and allelism analyses of hypernodulation soybean mutants from two gentetic backgrounds, *Crop Sci.*, 36, 1153-1158

Witty, J.F. & Minchin, F.R. (1990) Oxygen diffusion in the legume root nodule. In *Nitrogen Fixation: Achievements and Objectives*, ed. Gresshoff, P.M., Roth, L.E., Stacy, G. and Newton, W.E., Chapman and Hal, New York, London, 285-292

Yashima, H., Fujikake, H., Sato, T., Tewari, K., Ohtake, N., Sueyoshi, K., & Ohyama, T. (2003). Systemic and local effects of long term application of nitrate on nodule growth and N_2 fixation in soybean (*Glycine max* (L.) Merr.). *Soil Sci. Plant Nutr.*, 21, 981-990

Yashima, H., Fujikake, H., Yamazaki, A., Ito, S., Sato, T., Tewari, K., Ohtake, N., Sueyoshi, K., Takahashi, Y. & Ohyama, T. (2005). Long-term effect of nitrate application from lower part of roots on nodulation and N₂ fixation in upper part of roots of soybean (*Glycine max* (L.) Merr.) in two-layered pot experiment. *Soil Sci. Plant Nutr.*, 21, 981-990

5

Causes and Consequences of the Expansion of Soybean in Argentina

S. Calvo[1], M. L. Salvador[1], S. Giancola[2], G. Iturrioz[2]
M. Covacevich[2] and , D. Iglesias[2]
[1]National University of Córdoba
[2]National Institute of Agricultural Technology –INTA
Argentina

1. Introduction

Soybean production in Argentina started in the seventies "as a productive option to provide proteins for animal feeding. This was fostered by the National Institute of Agricultural Technology (INTA) at a national level and by the Argentine Institute for the Oilseeds Development (IADO) , currently closed" (ACSOJA, 2009). In spite of this, the soybean development was very slow until the mid 90s, when two technological milestones occurred: no-till farming and the inclusion of transgenic seeds.

The development of this chain is shown in its growing competitive power. Indeed, Argentina has kept and increased its presence in the world market, being the third exporting country of soybean (13% of the total exports). It is also the first world exporting country of soybean oil and meal (55% and 50% share of the total exported). Finally, the export goods of the soybean chain represent the first exportation item of the Argentine economy (2010/2011).

This may be explained by four factors: a) the agro ecologic suitability, b) the constant improvement and technological innovation, c) the most modern and of major scope crushing facilities for soybean of the world, d) the high exportable surplus due to low domestic consumption.

In the last ten years, the development of this chain not only has favored the members of the chain –producers, manufacturers, exporters, merchants- but also the National State, which by means of the implementation of export duties[1] obtains strong revenues.

In spite of the favorable setting in which the soybean chain appears both, at a national and international level, this crop is questioned by several agents who analize how the soybean spreading is shifting the agricultural frontier. This expansion in turn is causing the substitution of agricultural and stockbreeding products, the use of not suitable lands for agriculture, the extinction of small-size producers and environmental damage among other consequences.

Therefore, the goal of this work is to describe and analize the evolution of the soybean chain in Argentina through indicators and explicative factors[2] referred to the primary, industrial

---

[1] Up to March 2010, they amount to 35% for grain and 32% for soybean oil and meal·

and exporta sectors, since the mid seventies up to the present time, with special emphasis on the last thirty years. The aim is to outline the probable future context that the direct and indirect agents connected to the soybean complex will face.

2. Primary production

At the beginning of the seventies, the soybean cultivated area reached 30,470 hectares with a total output of 26,800 tons. The low importance of this crop may be explained by the lack of attractive markets, the low international prices and the national trade policy, which strongly restricted the exportation of grain and byproducts (Civitaresi and Granato, 2003).

The expansion of the cultivated area and yields started during the seventies. The cultivated area increased 6,792 % between the 1969/1970 and the 1979/1980 seasons and 143 % between 1979/1980 and 1989/1990 whereas the production increased 12,959 % and 206 % considering the same periods. Likewise, the soybean yields increased 32.7 % between 1970/1971 (1,024 kg/hectare) and 1989/1990 (2,157 kg/ha). This outstanding growth was due, among other endogenous causes, to technological changes and its fast adoption shown in the improvement of agronomic management, agricultural mechanization and agrochemicals application (Civitaresi and Granato, 2003). In addition, the double crop wheat-soybean[3] began to develop.

Among the exogenous causes that explain the growth in this twenty years (1970/1990), highlights the increase of international prices in the mid seventies, reduction of export duties, elimination of exportation ban (1978) and modification of the policy carried out by the European Economic Community (EEC) nowadays European Union (EU). In the 70s, the EEC negotiated the entry of soybean and its byproducts with zero tariff according to a re-structural procedure of its Common Agricultural Policy in the framework of the Dillon Round of the General Agreement Tariff and Trade (GATT) (1960/1961). The strong expansion of its milk production and the subsequent need to feed its cattle urged the strong demand of vegetable proteins. The EEC tried -unsuccessfully- to cover its needs with its own cereal production by restricting the access to imported cereals (as the wheat bran from Argentina), but it could not restrict the entry of soybean, which was protected by the commercial concession in the framework of the GATT[4] (Albin and Paz, 2003).

In the last twenty years, the process of expansion of the soybean cultivated area has continued uninterrupted. The cultivated area increased 72.3 % between 1989/1990 and 1999/00 and 108.7 % between 1999/00 and the last season 2009/2010. In addition, the soybean production increased 88.2 % and 162 % for the same periods.

In Figure 1 it can be seen the expansion of the cultivated area and of the soybean production[5] as well as the leap in both indicators since the year 1996.

---

[2] It will be applied the definition by Civitaresi and Granato (2003) which classifies the factors into exogenous: that depend on macro economic policies and on international commerce, and endogenous: that derive from the strategies carried out by the protagonists of the different links of the chain.

[3] The short cycle wheat allows the sowing of wheat-soybean in the same agricultural cycle and in the same area

[4] In the eighties, it tried to "under strengthen" the soybean and its derivatives: this situation led to serious confrontations with the U.S.A.

[5] The decrease in production in 2008 and part of 2009 was due mainly to a drought and partly to the conflict between the agricultural sector and the Government. The conflict was originated by the implementation of "mobile export duties" by the Economy Minister (the duties varied according to the

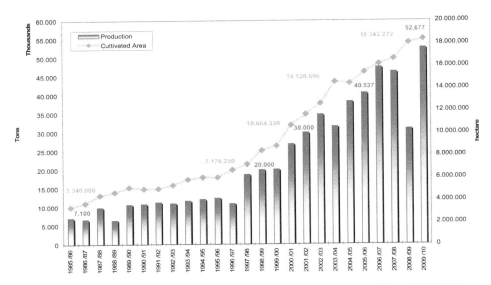

Fig. 1. Argentina soybean. Production and Cultivated Area. Years 1885/86 to 2009/2010.

Specifically in the nineties, the following endogenous factors can be featured as causes of the expansion: the genetic modification in several crops –among them the events in the soybean seed and consequently the start of the transgenic technology spreading at an international level-, and the intensification in the use of machinery and supplies, mainly the "glyphosate"[6] herbicide. In addition, a fundamental change in the plowing system took place: no-till cropping, which deserves a special paragraph.

"No-till cropping started to gain importance in the Argentine agriculture in the late eighties since in many of the most important areas of the *Pampeana* region, the cumulative effects of the soil erosion -resulting from the "agricultural cultural process" on the basis of traditional plowing practices– began to show negative operative results in the exploitation. This effect on yields, and through them, on the economic feasibility of agriculture, together with the fact that, as a result from the deregulation and opening processes of the economy, the availability of appropriate farming machinery increased, plus the reduction in direct costs (as a result of the elimination of the plowing tasks), provided an outstanding basis for the spreading of no-till cropping, which in turn had the objective of recovering, at least some of the lost productivity" (Trigo and Cap, 2006).

Fuethermore, the synergy between glyphosate tolerant soybean tolerant (RR-soybean) and the non-till system –which shortens the required time between the wheat harvest and the soybean planting- makes possible the successful use of short cycle soybean varieties as second crop and allows a double-cropping wheat-soybeans in zones where before was not possible from an agronomic point of view. According to data provided by the Argentine No Till Farmers Association (AAPRESID), at the beginning of the nineties,   the area of soybean

---

variations in the international price of the soybean). This duties were enforced from March 2008 to July 2008 (they were derogated by a legislative decision).

[6] In the 2009/2010 season, more than 200 millon liters of glyphosate were applied in all the soybean sown area in Argentina, whereas in 1996 13 million liters were used

under no-till cropping system reached 6%, whereas in the 2001/02 season it was 74%, in 2006/07 it accounted to 85% and in 2009/10 it reached 88 %.

Analyzing the exogenous factors that fostered the development of soybean in the nineties, highlights the changes in the economic policy: tariff reduction for machinery and inputs imports, reduction and/or elimination of export duties for products of the soybean chain, investments in ports, deregulation of markets (elimination of Meat and Grain Boards) and privatizations (ports, railways) which lessen the cost of production and favor the acquisition of technology.

At an international level, relevant facts such as average international prices lower than in the previous decade , the increase of commercial barriers in the rest of the world (in spite of the final Agreement of the GATT´s Uruguay Round) have not been auspicious for the soybean chain. In this respect, the only favorable fact was the increase in the demand of grain by the Brazilian oil industry, in the frame of the MERCOSUR (Civitaresi and Granato, 2003).

Going through the last decade, the increase in the soybean cultivated area[7] is due both to the incorporation of new lands and to the substitution of other crops. The agricultural expansion towards the north of Argentina[8] is shown by a 70 % increase of the cultivated area between the last two National Agricultural Census (CNA 1988 and CNA 2002[9]), although this is a marginal area for this crop. 66 % of the lands transferred to agriculture in the north of Argentina was occupied by soybean (Giancola et al., 2010).

According to Shvarzer and Tavosnanska (2007), the increase in the production for the last ten years might have several causes. One of them is the increase in international prices because of the growing world demand for soyean oil and meal. Another cause is the technological change, which involved a reduction of related costs. Likewise, between the years 2000 and 2010 it is worth mentioning: the intensification of the technological package, scale incrementes in the productive core, the suitability of the soil, the existing road infrastructure and the increase of the rainfall (Giancola et al., 2010).

Complementing the previous considerations, among other fundamental causes of the soybean expansion is worth mentioning: higher financial income, lower complexity in the operation and lower risk than other crops, and farmers' knowhow, together with the use of their own seed (INTA, 2009).

As related to the topic of seeds and specially about the use of the varieties of genetically modified soybean[10] (GM), the data provided by ARGENBIO (a non governmental organization created to spread information about biotechnology) shows that in the 1996/1997 season, the area occupied with GM soybean was 6% and quickly reached 94% in 2001/2002 to cover 98% in 2004/2005.

---

[7] At present (2009/10 season) soybean occupies 18,343,272 hectares with a production of 52,677,371 tons (CIARA, 2011).

[8] The traditional region for the production of soybean is called "pampeana" and it is made up by the provinces of Córdoba, Buenos Aires, Santa Fe, Entre Ríos and La Pampa.

[9] At a primary level, 75% of the production comes from 40% of the productive units (100,000 units). The average size of the units is 170 hectares.

[10] The gene resistant to the glyphosate herbicide was initially owned by Monsanto in the United States, which granted a license to Asgrow, then this company was bought by Nidera, which introduced it in Argentina. Afterwards, when Monsanto patented the product abroad, it had already been freed by third parties to be sold in Argentina (Quaim and Traxler, 2002). This situation started a controversy between Monsanto and the Argentine government, in which the company has claimed in order to obtain royalties.

Even in Argentina, these adoption was faster than others, as for example hybrid corn or wheat with Mexican germplasma. Hybrid corn lasted 18 years to reach 70% of the acceptance that nowadays GM corn holds, and Mexican wheat reached the same percentages of adoption as soybean has nowadays (more than 90% of the market) only after 16 years (López, 2006, as cited in Trigo and Cap, 2006).

Another factor that favored the technological change is related to inputs supply for the soybean crop (Bisang, 2003) which has the following characteristics: a) technological packages: the transgenic seed -RR Soybean- works as a connector of a joint supply that includes: glyphosate, pre-emerging herbicides, insecticides and fertilizers; b) mergers and alliances with companies that offer such technological packages[11]; c) the packages of supplies offered in service centers include advice about usage techniques.

Finally, soybean share of total cultivated area and grain production of the country allows to visualize the strong soybean expansion in Argentina (Figure 2). According to production data, soybean represented 0.2% of the total cultivated area[12] (14 million hectares) in the 1969/1970 season. In the seventies, the soybean cultivated area occupied 3.1 % of the total sown, in the eighties 19%, in the nineties 33% and in the first ten years of the 21st century soybean occupied 54% (more than 18 million hectares) of the total (28 million hectares) implanted.

Regarding the importance of soybean production, and considering the total produced by the five main grains, the participation of soybean was 3.7% on total production in the seventies, 18% in the eighties, 30% in the nineties and 50% in the first ten years of the 21st century (CIARA, 2011). In the 2009/2010 season, the soybean production represented 60% of the total production whereas it was 0.12 % in the 1969/70 season.

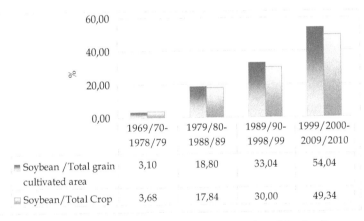

	1969/70-1978/79	1979/80-1988/89	1989/90-1998/99	1999/2000-2009/2010
■ Soybean /Total grain cultivated area	3,10	18,80	33,04	54,04
▢ Soybean/Total Crop	3,68	17,84	30,00	49,34

Fig. 2. Argentina. Soybean. Production share of the total produced by the five crops and soybean cultivated area shares of total cultivated area. Periods: 1969/70 – 1978/79, 1979/80 – 1979/80 – 1988/89, 1989/90 – 1998/99 and 1999/2000 – 2009/10.

---

[11] Monsanto (an company originally dedicated to chemistry, which in the70s expanded to pharmaceutical chemistry) is associated with Dekalb and Cargill; the use of these seeds (specially RR Soybean from Monsanto) requires the application of glyphosate, offered by the same company with the brand "Roundup".

[12] Taking into account the five main crops: wheat, corn, sunflower, sorghum, soybean.

With respect to the geographical location, the soybean cultivated area is situated in 15 (fifteen) productive provinces. However, 85% of the cultivated area is concentrated on the traditional "pampeanas" provinces that generates 88% of the national production (Córdoba, Buenos Aires and Santa Fe provide 78% of the total production) whereas 15% of the remaining area corresponds to Northwester and Northeaster provinces. Map 1 shows the shift of the agricultural frontier[13].

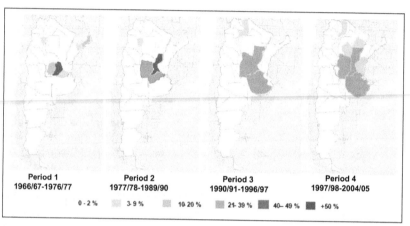

Map 1. Evolution of the space distribution of the soybean cultivated area (Brieva, 2006).

Considering the yields, Figure 3 shows that they have increased continuously, from 1.4 tons/ha in the seventies, to 2.6 tons/ha in the first decade of 2000.

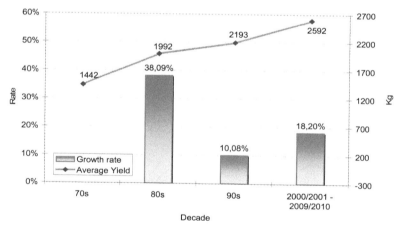

Fig. 3. Argentina Soybean. Average Yields (kg/hectare) and growth rate (%). 70, 80, 90 decades and 2000/01 – 2009/10 period.

---

[13] As mentioned, the shift of the agricultural frontier has generated the substitution of crops (sorghums, corn, cotton among others) and cattle livestock for soybean as well as the spreading of soybean over the native forest.

3. Industry and state policies

In the stage of first industrialization or milling, the productions of raw oil and meal are technologically associated, and therefore they are produced in the same industrial plants. Most part of the grain production in Argentina is oriented to milling, with a percentage that varies from 76% to 80% of the total production; approximately 18% of soybean beans is exported without processing and the rest is used for direct consumption, as seeds for sowing or for other purposes (bread, cookies, snacks, peanut butter, bird food, etc).

The industrialization process begins with the milling and extraction of oil. After going through drying - to remove humidity – and cleaning processes, the grain is broken and pressed in small sheets, which being transformed into a dough, move on to the extraction process. The remaining dough, after being dried and toasted, forms the protein meal used in the manufacture of animal feed. The gum can be used in the production of soybean lecithin or can be added to the meal in order to obtain different of protein tenors. Through a hydrogenation process, the partially refined oil can be transformed into margarine, mayonnaise and vegetable fats. Some companies integrate vertically these industrial stages.

As other grains with a low oil component, soybean oil is extracted with hexane. The solvent –a petroleum product- impregnated with oil is then separated by evaporation, and goes through a gum elimination system (degummed) to reach the stage of raw oil (Giancola *et al.*, 2010). A soybean yields 19% oil, 73% meal, 7% shell and 1% others (ashes, etc.), these values may vary depending on the drying and de-shelling degrees[14].

The manufactured products[15] are distinguished in two segments: **commodities** (for exportation and domestic market), which include raw and refined oil in bulk and meal for animals, and **speciallities** of higher added value for the final consumption of food and other uses. Some companies are present in both segments. The refined oil, apart from its direct consumption (pure or in blends) has several uses, such as margarine, mayonnaise, non edible intermediate products (candles, cosmetics, soap, paint and fine chemicals, animal food, soybean lecithin), hydrogenated vegetable fats (used in cookies, bread, ice-cream, etc), soybean derivatives for human consumption (Gutman and Lavarello, 2003). Regarding non edible uses and oriented to the chemical industry, oil and vegetable fats have similar characteristics to the petroleum ones and can be used for plastics, adhesives, solvents, lubricants, etc.

Another derived product is biodiesel. Compared to other traditional hydrocarbons, biodiesel has several advantages: it comes from a renewable product and has little environmental impact, but it also has disadvantages, being its high cost the main limitation. Biodiesel installed capacity of production is to 2.43 million tons – reaching 3 million tons by the year 2011. In 2010, 1.9 million biodiesel tons were refined and almost 1.4 tons were exported. The 500,000 remaining tons were used in the domestic market for the diesel cut.

In April 2006, Act 26093 established the rules for the regulation and promotion[16] for Production and Sustainable Use of Bio Fuels for a period of 15 years. This Law establishes that the gasoline and diesel commercialized within Argentina shall be mixed in an oil

---

[14] 4 tons of flour are obtained from each ton of oil.
[15] The industrial processing generates byproducts that have more restricted but dynamic markets, many of which are not developed enough in Argentina.
[16] The Law gives encouragement for investment through fiscal incentives.

distillery or refinery with 5 % - as minimum – of bio ethanol and biodiesel, respectively from 01/01/2010 on. For the year 2010, the quantity of biodiesel necessary for the 5 %[17] cut was calculated in 625,000 tons, which imply 650,000 tons of oil.

Likewise, the biodiesel industry that is developing in Argentina has a strong orientation towards the foreign market[18]. In Argentina, there are major incentives for the biodiesel exportation as a result of the way in which export duties of the soybean complex are structured. In this respect, the biodiesel exportation must pay 20 % of export duties whereas soybean oil (its main supply) and soybean pay 32 % and 35 % respectively[19]. Thus, there is a gain of competitiveness for biodiesel that comes from the efficiency in the previous links of the chain. In addition, biodiesel obtains a reimbursement of 2.5 % upon the export price, with which the competitiveness reached by the export duties structure is reinforced.

Finally, it is worth mentioning that it is currently spreading the building of small extruding facilities in farms that produce soybean. This allows to obtain oil for its use as biodiesel and protein meal for cattle feeding, adding value to thier soybean production through horizontal and vertical integration

3.1 Milling capacity

Milling is the most important destination of the primary production of soybean. While in the eigthies 71 % of the grain production was assigned to milling, in the 90 decade this percentage increased to 77 % and reached 84 % in the first 2000 decade. In the 2008 and 2009 periods, there was a 20% fall in industrialization (30 million tons average) which can be explained mainly by the decrease in the grain production due to reasons pointed earlier. In the year 2010, the milling reached 34 million tons, similar values to those of the year 2007. This volume represented 98% of the soybean supply.

In the last years, the crushing capacity grew basically in the province of Santa Fe (at an average rate of 14.4 %) while in the provinces of Buenos Aires, Córdoba and Entre Ríos it remained practically constant (CIARA, 2011). In the year 2010, soybean crushing in Argentina reached 36,824,628 tons, which implies 92% of the total milling of the country (39,898,017 tons[20]. In 1986, the soybean milling share of the total milling was 50%. (CIARA, 2011).

The increasing crushing capacity in Argentina is also reflected in the world market. In the last two decades, the world milling has grown at an annual rate of 4.5 %. In the same way, the soybean world production is destined mainly to milling (98 %) reaching 200 million tons (2009/2010)(USDA, 2010). The main countries processing soybean are also the main grain producers: U.S.A. Brazil, Argentina, China. Currently, this latter country – which specially has the goal to increase the soybean added value – contributes 21 % of the world milling. This is the reason of the strong grain importation from overseas.

---

[17] The national government established that the cut shall be 7 % since 2011.

[18] Nine plants have been authorized to export (2008), 5 of which come from investments made in the oil sector, the rest come from non agro-industrial sectors. The main refineries are: Renova, Louis Dreyfus, Unitec Bio, Patagonia Bioenergía, Ecofuel, Cargill and Aceitera General Deheza. CARBIO (Argentine Chamber of Biofuels, 2011).

[19] The export duties of grain and byproducts had been applied in previous times, as in the eighties (15 % and 41 % for grain) and nineties (3.5 % for grain).

[20] The crushing capacity in Argentina amounts to more than 57 million tons. (2010)

3.2 Oil and meal production

Soybean oil production reached 41 million tons, being the main producing countries: China, (10,3 million tons), U.S.A, Brazil and Argentina. Argentina produces 7,3 million tons (2009) and participates approximately with 18 % of the world production.

Argentina has 54 oil producing plants which are distributed in 8 provinces. Thirty-nine of these plants process soybean bean with a theoretic capacity of 160,000tons in 24 hours (CIARA, 2011). Fifty-six percent of the plants that concentrate 84 % of the processing theoretic processing capacity, operate in areas close to shipment centers in the Province of Santa Fe, since the production is oriented to exportation. The processing plants are supplied with soybean within a radius smaller than 300 km, which results in low freight cost. This closeness between primary production and transforming industry generates a major competitive advantage.

Analyzing Table 1, we can see that whereas in the 70 decade, the soybean oil production participated in 8.5% of total elaborated oil of the country, in the 2000/2004 period this participation increased to 71.88%. In 2009, soybean oil participation in Argentina reached 5,771,812 tons, thus participating 79% of the total elaborated oil (7,302,493 tons). Considering the geographical location, the province of Santa Fe elaborated more than 90% (2008/09) of soybean oil total production.

	Total Oil (T)	Soybean Oil (T)	Soybean/total (%)
Decade Averages			
1970-79	661.517	56.118	8,48
1980-89	1.735.284	571.359	32,93
1990-99	3.722.981	1.789.712	48,07
Quinquennial averages			
1985-89	2.239.749	842.542	37,62
1990-94	2.951.163	1.360.297	46,09
1995-99	4.494.799	2.219.126	49,37
2000-04	5.444.255	3.913.555	71,88

Table 1. Argentina. Total Oil and Soybean Oil Production (tons). Soybean Participation (%) in the Oil Total. Periods Selected.

The development of the Argentine oil industry was parallel to the growth of the soybean primary production. According to Obschatko (1997) and Civitaresi and Granato (2003) the main factors were: the increase[21] in the international prices of oil and meal, impelled by the increase in the world consumption and the fluent grain supply (raw material) -at an international level-, and the reduction of taxes for the exports of oil and the application of a lien differential for grain and oil –at national level. The latter measure was fundamental for the development of the industry, according to Obschatko (1997)- in order to prevent the industry from having problems for raw material supply, the National State

---

[21] In the nineties, international prices decreased due to the lower demand by Asian countries and higher presence of other substitute oils. All this took place in a context of strong market distortions due to protectionism mostly inin developed countries.

implemented an effective differential exchange rate for grain, oil and byproducts. Between 1976 and 1990, the differential of export duties between grain and milling products (highly praised) varied between 5.9 and 13.6 %. This measure favored the exportation of oil and meal by reducing the cost of the raw material with respect to the international market prices.

At an industrial level, the investment made in this sector as regards technology was complemented with the availability of technology from developed countries- that is the case of the Dutch enterprise (De Smet), which introduced in Argentina the technology of oil extraction by solvent. Likewise, the companies of the industrial link increased the production scale: they specialized in soybean and sunflower; they re-located in the province of Santa Fe –thus combining the closeness to the exportation ports (Rosario) and the sources of raw material supply. Finally, industry integrated forwards by building dock[22] and storage facilities. This process was intensified in the nineties. At a National State level, the tax pressure on the marketing sectors was reduced, the services related to trading activity were deregulated, public ports were privatized and new ports were built. In addition, there were investments for the sweeping and beaconing of the *Paraná* and *Río de la Plata* fluvial corridor[23] (Civitaresi and Granato, 2003).

Soybean processing is concentrated in six companies: Bunge, Cargill, Vicentín, Molinos Río de la Plata, Dreyfus, Aceitera General Deheza, which control more than 87 % of the total capacity. These companies have a modern and efficient port infrastructure of their own and an extended supply net spread throughout the entire country[24]. The oilseeds industry employ 10,000 people. In Table 2, it is identified the number of plants that each of the companies owns as well as the annual processing capacity.

Company	Number of Facilities	Total capacity		Largest capacity Facility (ton/year)
		Ton/year	%	
Bunge	5	8.220.000	17.8	2.550.000
Cargill	4	7.710.000	16.7	3.900.000
Vicentín	3	6.555.000	14.2	3.000.000
Molinos Río de la Plata	3	6.195.000	13.4	3.600.000
Dreyfus	2	6.000.000	13	3.600.000
Aceitera Gral. Deheza	4	5.700.000	12.3	2.550.000
Buyati	2	1.440.000	3.1	1.005.000
Nidera	2	1.260.000	2.7	660.000
OMHSA	2	705.000	1.5	405.000
Others	20	2.467.500	5.3	
Total	47	46.252.500	100	3.900.000

Source: Schvarzer and Tavosnasnka (2007)

Table 2. Argentina. Oil Factories by Company.

---

[22] In 1977, private companies were allowed to set up fluvial ports.
[23] This corridor is fundamental for the transportation of the grain production from Córdoba and Santa Fe, first and second producing provinces of soybeans.
[24] As regards the production of refined oil, 5 companies have 80 % of the production, 4 companies have 80 % of the margarine production and three companies have 80 % of the mayonnaise production.

At a world level, and according to USDA data, the production of soybean meal was approximately 177 million tons (2009/10), being the main producers: China, (25.7% of the total produced), U.S.A., Argentina (17% of the world production) and Brazil. These four countries concentrate 78% of total production (USDA, 2010).

The soybean meal share of total manufactured meal in Argentina has been higher than oil. Thus, whereas soybean meal was 20.87% of the total output in the seventies, this percentage reached 91.1% in the 2000/04 period (Table 3). In 2009, the soybean meal production was approximately 23 million tons from a total of 25,520,000 tons of manufactured meal. As in the case of oil Santa Fe province produces 93% of total soybean meal (2009).

Period	Total meal production	Soybean meal production	% soybean/ total meal
Decade average			
1970-79	1.268.956	264.807	20,87
1980-89	4.224.577	2.665.821	63,10
1990-99	10.383.969	8.124.909	78,24
Quinquennial average			
1985-89	5.674.917	3.916.431	69,01
1990-94	8.248.517	6.327.119	76,71
1995-99	12.519.422	9.922.700	79,26
2000-04	18.202.340	16.584.419	91,11

Source: CIARA (2010)

Table 3. Argentina. Total Output of meal (tons) and of soybean meal and Participation (%) in the Total Output of meal. Periods selected.

4. International trade

Regarding soybean complex, Argentina exports mainly processed products: protein meal for animal feeding and oil for human consumption. Soybean exportation began its expansion due to the demand from China, which intensified its process of importation substitution (oil).

4.1 Soybean

The world production of soybean in the 2010/2011 season reached 255.5 million tons according to USDA. The main producing countries, U.S.A. (35 %), Brazil (26 %) and Argentina (20 %), participate in 82 % of the world production. They are followed by China, with a participation of 6 % although its presence in the world market has the role of importer because of its high domestic consumption.

For a long period of time soybean main demand came from the EU. Since 1998/99, China has incremented its demand for domestic processing and since the 2004/05 harvest China turned into the main world importer of the grain (60 %) followed by the EU (14.6 %).

As related world suppliers, U.S.A, Brazil and Argentina concentrate 89 % of the total exported (98 million tons) (2009/2010). Argentina occupies the third place as soybean exporter in the world with 14,11 % of the total[25] (see Figure 6). Taking into account the six main soybean exporting countries (U.S.A, Brazil, Argentina, Paraguay, China, Canada, Bolivia), three of them – Brazil, Argentina and Paraguay – which make up the MERCOSUR-represent half of the world exportations. This points out the importance of South America in the soybean bean production and exportation.

As seen in Figure 4, the soybean exportation has had a positive trend fostered by the increasing international prices. This indicator -grain exports-[26] shows the higher competitiveness of Argentina in the international market.

The main demand of Argentine soybean comes from China. Thus, 78 % of the exported grain had that destination. Argentina also exports to Iran, Thailand and Egypt, and in South America to Chile and Peru.

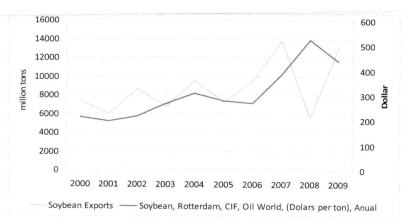

Fig. 4. Soybean. Argentine exportation (million tons) and price (U$S/ton) (CIF, Rotterdam) 2000/2009.

Argentina has also imported soybean, especially from Paraguay (98 %) (Figure 5). The need to import soybean can be explained since the processing capacity of the oil industry surpasses the local production of soybean and other oilseeds available for industrialization. The strong increase in soybean importations since 2006 can be explained by the Decree 2147/06, which was the result of a wide negotiation between the industrial-exporting sector and the national government. This decree allows transitory importation of goods that will be subject to a process of industrial improvement. Decree 2147/06 incorporates a tax benefit upon the imported supply which allowed the exporting companies to pay export duties omly on the added value of the imported bean. Since April 2009, the Argentine Production Ministry excluded[27] the soybean from the tax benefits of temporary importations. This regulation was based on the grounds that it would promote the increase of the soybean demand in the domestic market and diminish the stock of grain stored by producers in their own farms. (Giancola et al., 2010).

---

[25] U.S.A participates with 44 % and Brazil with 32 % of the total world exported of soybean.
[26] The decrease in the grain exportation is related to the drop in the production due to a drought (2008 and part of 2009).
[27] That is why no importations have been made since 2009 up to the present.

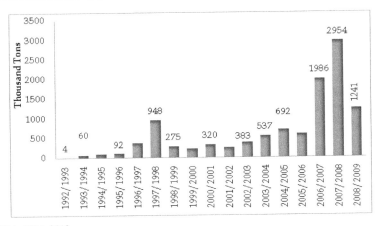

Source: USDA PSD ,2010

Fig. 5. Argentina. Importation of Soybean Bean (tons). 1992/93 – 2008/2009.

4.2 Soybean oil

The soybean oil exportation is concentrated in few countries whereas the importation is diversed. So, the first five importing countries (China, India, UE-27, Iran, Morocco) have 52% of the total imports (2010-2011).

Argentina is the main oil exporting country[28]. With 49% to 56% of the world total oil exports, followed by Brazil (16%) and the United States (13%). Considering Paraguay, the MERCOSUR is regarded as the main exporting soybean oil block. 96% of the oil exported by Argentina is raw and degummed. The refined oil incidence is very low.

The following table (4) shows Argentine apparent consumption. The data show that most of the oil production is destined to the exports.

In Figure 6 it is shown the Argentina share of total world soybean oil exported. Its main participation was in the 2005-2006 season (57%). Then, this participation decreased until it reached 49% of the 2009/2010 season. According to USDA estimates, in 2010/2011 season the Argentine exports would reach historical values (54.6%).

Year	2004	2005	2006	2007	2008	2009
Inicial stock	242.326	134.024	239.000	217.578	302.083	303.968
Production	4.569.718	5.395.724	6.161.214	6.962.675	6.024.101	5.771.812
Availability	4.812.044	5.529.718	6.400.214	7.180.253	6.326.184	6.075.780
Exports	4.588.120	4.964.180	6.086.290	6.637.770	5.125.480	4.660.400
Final stock	134.024	239.000	217.578	302.083	303.968	287.668
Apparent consumption	89.900	326.538	96.346	240.400	896.736	1.127.712

Table 4. Argentina. Apparent consumption of soybean oil.

[28] It is important to point out that since 2007, Argentine began to export biodiesel obtained from soybean.

Asian countries concentrate most of the Argentine oil demand due to the importance of their growing economy and to the fact that these countries had consumption levels per capita lower than the world average. China (45%)[29], India (14%), Bangladesh (6%), Egypt (5%), Peru (5%) are the most relevant destinations of the exported oil by Argentina (2010). Seventy eigh percent of the oil imports and 22% of the soybean imports of China have their origin in Argentina. This shows the strategic importance of soybean and its byproducts in the commercial relationship between China and Argentina.

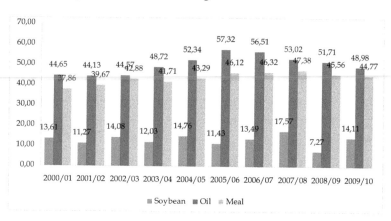

Fig. 6. Soybean. Argentina share of the world grain, oil and meal exports. (%) 2000/01-2009/10.

Finally , after a study of the soybean oil exports per company (Table 5) it comes out that the soybean exports market structure is an oligopoly[30], with foreign capital predomination (transnational companies) (Eumercopol, 2007).

Company	Soybean oil share of total oil exported (%)	Company	Soybean oil share of total oil exported (%)
Cargill	28,42	Nidera Arg.	3,12
Bunge Argentina	18,15	Aceitera C.A.	2,65
Aceitera Gral. Deheza	14,80	Oleag. Moreno	1,44
Dreyfus	10,91	Cia. Argentina de Granos	1,39
Mol. Río de la Plata	10,36		
Vicentin	3,31	A.F.A.	1,22

Source: CIARA (2011)

Table 5. Soybean Argentina. Soybean oil exporting companies. Total export share 2008.

---

[29] Due to the commercial restrictions imposed by China to the Argentine oil, the exports were drifted to the markets in India, Bangladesh and Iran at a lower price.
[30] The exports participation of the first eight companies (C8) is: 81% for grain, 87% for oil and 97%.for soybean meal.

4.3 Soybean meal

As in oil, Argentina is the main exporting country with 29 million tons over an exporting total of 59.4 million (Figure 7). Thus, Argentina keps -in a stable way- 44--46% (see Figure 6) of the total meal exports, followed by Brazil (23%) and the United States (14%). According to USDA for 2010/11 season, Argentina share in soybean meal export would reach 49.3%.

Argentina exports almost all its soybean meal production (between 93% and 97%), mainly to the UE-27 countries for animal feeding. The European problem with bovine spongiform encephalopathy, known as the "mad cow disease", put an end to the cattle feeding with animal proteins and intensified the demand of substitute products like the soybean meal.

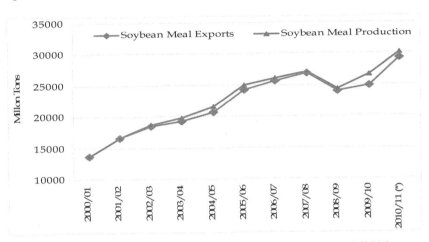

Fig. 7. Argentina. Soybean meal. Production and exports (2000/2001-2010/2011).

If the soybean meal exports by each company is examined (Table 6). The first five companies participate with 78.5% of the total national exports. It can be seen that the first four meal exporting companies are the same as the first four oil exporting companies being Cargill the first (28.4% in oil and 23% in meal, 2008).

Company	Share of total oil export (%)	Company	Share of total oil export (%)
Cargill	28,42	Vicentin	3,31
Bunge Argentina	18,15	Nidera Arg.	3,12
Aceitera Gral. Deheza	14,80	Aceitera C.A.	2,65
Dreyfus	10,91	Oleag. Moreno	1,44
Mol. Río de la Plata	10,36	Cia. Argentina de Granos	1,39

Source, CIARA (2009)

Table 6. Soybean Argentina. Exporting soybean meal companies .Participation in the total exported .2008.

5. Domestic commercialization, storage and logistics and transportation services

5.1 Comercialization. Main characteristics

Ninety five percent of the soybean harvest goes from farmers to country grain elevators (private or local cooperatives). The main final destiny is the crushing industry (oil and meal). The second destiny in importance are the exporters. The remaining activities have a small participation in the grain final destiny. Among these activities the animal feed producers and the grain purchase for own consumption (cattle producers) can be found.

The commercialization stages are divided according to the offerers, been primary operations when the seller of the goods is the producer; secondary operations when the goods are sold by any of the intermediary agents to exporters or to the industry, and finally, terciary operations which are those where exportation is involved.

In the primary stage, the "producers" are the ones who make up the supply : they can be land owners, tenants, contractors, pool producers, agricultural investments common funds, and the demand are usually country grian elevators, first grade cooperatives, livestock producers (poultry, pork, feed lot), exchange dealers[31], industries and expoterers with vertical integration strategies.

The secondary stage is focused on country grain eleators, local cooperatives, inputs providers and exchange dealers as well as exporters, industrials and livestock producers.

Brokers and institutions like the Stock Exchange, Cereal Boards and Arbitration Chambers can also take part in any of these stages. They –boards and chambers- are Arbitration Courts in case there is any kind of controversy. They set reference prices and have grain analysis laboratories. In Argentina, there are four Cereal Boards: Buenos Aires, Córdoba, Bahía Blanca and Entre Ríos.

As regards the transaction steps, the most representative ones can be divided between those made with merchandise readily available or not. The most common payment ways when the goods delivery is effectively done can be: cash on delivery, cash with insured delivery quota, cash with uncertain delivery, cash on deposit certificates or a price to be fixed. When the goods are to be delivered in the future, the payment ways can be: advance payment, exchange (with other goods), and forward payments (with a contract).

In Argentina, there are also two exchange markets where Futures and Options contracts on soybean can be made: "Rosario Futures Exchange (Rofex)" and "*Mercado Término de Buenos Aires (MATBA)*" where soybean oil contracts can be found. Although the use of these instruments is growing; it is not widely spread yet, specially among the small producers. Besides, successive economic crisis and interventions in the agricultural markets have slowed down its development. It is important to point out that this alternative is use to secure prices rather than to deliver the goods. The use of these markets is around the 28% of the total average production in the last five years (2005-2009).

The operators that take part in grain commerce have been classified according to their function into 17 categories by the National Office of Agroindustrial Commerce and Control (ONCCA, 2006). The ones involved in soybean markets are: country grain eleators (2685), grain conditioners (62), industrials (58), industrial-selector (35), laboratory (25), public scales (21), Futures Market (3), consigner without facilities (2), broker (359), importer (27), exporter (356), exchange agent (665) and delivery agent (11).

---

[31] Exchange dealers receive soybean as a form of payment for goods in general and for leasing.

Different trading channels and modalities combined with the large number of operators makes it possible to trade soybean in a very extended way. But since it's not compulsory to register the operations modality, no statistics reflect in a reliable way which the most chosen channels are. Anyway, these modalities are in general incidental and depend on the market situation.

5.2 Storage
The expansion of the storage capacity in Argentina was historically related to the volume and composition of the country grain production and also to the politicies officially developed in relation to the way of trading. In the last years, an expansion of the storage capacity is observed. This expansion had its origin, among other factors, in the sustained increase of grain production, which implied a growing need of places to keep this production. (Lopez, 2008)

	Tons	%
Grain storage companies-Cooperatives	38.204.066	53,86
Producer (fija)	15.900.000	22,42
Oil mills	7.655.511	10,79
Exports-Ports	4.759.119	6,71
Flour Mills	2.552.024	3,60
Animal Feed Proucers	785.947	1,11
Rice Mills	555.827	0,78
Conditioner	517.450	0,73
Permanent facilities	**70.929.944**	**100**
Silo bag	30.000.000	
Total	**100.929.944**	

Table 7. Argentine Storage Capacity

Producers kept the grain in the farm (when possible) or storage them in local cooperatives and country grain elevators or in the silos found in ports or factories (Table 7). National storage capacity in 2007 was nearly of 71 million tons. It must be mentioned that there's no official information in this area so the growth rate considered was similar to the one observed in former periods. (Lopez, 2008)[32]

5.2.1 The silo bag as a storage alternative
Between 1999/2000 silo bag[33] storage technique was introduced (Picture 1). It is a complementary capacity not only at a producer level but also in the case of some country grain elevators (FAO-SAGPYA, 2004). This technique consists of storaging the grains in hermetic plastic bags where the breathing process of the grain biotic components (grains, fungus, insects, etc.) consume the oxygen generating in turn carbone dioxide. This new

---

[32] It is important to remember information about the storage capacity in Argentina is scarce and in some cases of uncertain reliability (Lopez, 2008).
[33] The relationship installed capacity/silo bag was 2,66/1, 30% of total storage capacity. See Table 7.

atmosphere rich in carbone dioxide and poor in oxygen suppress, inactivate or reduce the reproduction capacity and/or fungus and insects development and also the proper grain activity which in turn makes it easier to store them (Casini, 2002).

Picture 1. Silo Bag

The spread of the use of this technology introduced new possibilities (it increased the retainining grain autonomy at a low cost) -although with certain limitations[34]- and also generated an additional service circuit formed by the bagging and unbagging offerers, as weel as the production of stuffing machinery (Bisang and Sztulwark, 2007). The increase observed in the use of silo bags, is an expression of a bigger storage capacity needed.

The use of this way of storing among the producers and country grain elevators allows them[35] to keep grain and make deliveries, avoiding high freight costs during the harvest time. If Soybean is stored in a dry and clean way, it can be kept between 4 and 12 months in good conditions. This allows to improve income up to 15.5 % (average, if sold in the month of January)(Ghida Daza , 2002).

5.3 Logistics and transportation services

Argentina has lower costs as regards production for most of the agricultural products where it has an important share of the global market, but it has higher costs as regards the commercialization than its most important competitors: Brazil and the United States (Dohlman et al., 2001; Tavarez 2004; Nardi and Davis, 2006). The higher commercialization costs are due to, in the case of soybean, higher transport, storage and exportation tariff (Nardi and Davis, 2006). Eighty four percent of the grain production is taken to the

---

[34] Possibility of bag breaking and subsequent loss of stored grain quality.
[35] Other advantages would be: to activate the harvest when the production is stored in the same productive land and to obtain credits on the stored grain (Warrant)

exporting ports by truck[36], 14,5% by rail and 1,5% by barges. Grain exports are made approximately 90% by ship, 7% by truck and the rest by rail and barge.

According to data from the Stock Exchange of Rosario, the Gran Rosario area (province of Sta Fe) is the one with the highest growth as regards land cargo transportation in Argentina in the last decade with annual volumes of around 60 million tons of grain products; 8 million tons of these annual volumes come by rail[37] and 52 millons by truck. This implies that during the year an average of 5.000 trucks per day are concentrated. This number increases during the harvest months. (Giancola *et al.*, 2010).

Although there is a tariff difference between the truck and the rest of the means of transport (truck versus barge 3.25 to 1, truck versus rail[38] 2.5 to 1 (Commercial Infrastructure Area) (SAGPYA, 2007), the strong participation of the truck for domestic freight is due to its speed and flexibility to adapt to the resources and conditioning structures. The storage centers are generally located in the productive areas or within a radius of 20 km and at an average distance of 300 km to the industrial centers and port terminals[39]. The national road infrastructure has a total extension of approximately 38.000km.

Motor carrier: The trucking fleet is approximately 400.000 units. For cereal and oilseeds transport there maybe 5,000 special units, but there are around 60,000 general cargo units adaptable to this need. The average cargo capacity of a truck is 28 tons. The average age of these trucks is around 25 years, in an atomized system of car properties.

Railway transport: One of the limitations to its development as a mean of transportation is that more than 1,700 establishments which work as storage centers (83% of the total) does not have rail access (ONCCA, 2006). The present rail granted companies which are now giving cargo services are: ALL Central, ALL Mesopotamico, Ferro Expreso Pampeano, Belgrano Cargas and Nuevo Central Argentino (NCA). (Giancola *et al.*, 2010). Each company attends a different region, and all reach the Ports of Rosario. Regarding the operating capacity for the grain transport and its derived products of all the net rails, there are nowadays nearly 6.500 wagons (40 tons in each grain wagon depending on the gauge) between solid and liquid cargo and a totalnet of approximately 28,000 km (SAGPYA, 2007).

Fluvial/Maritime transport: There are 40 Argentine port terminals which have the capacity to deliver grain, oil and protein meal in bulk. Eight of these are maritime and they are located in the province of Buenos Aires. The maritime terminals deliver 19%of the cargo of this kind of products. The other 24 shipment terminals are fluvial, and from them 81% of the grain, vegetable oil and protein meal is delivered.

Most of the soybean complex products are exported from the Paraná River, in the north and south of the city of Rosario, from San Martin Port to Arroyo Seco; this section covers 70 km.

---

[36] Compared to Brazil and the United States, Argentina is one of the exporting countries which makes the biggest use of the truck due to the average distance between the primary productive centers and the industrial processing .

[37] With the use of the motor carrier, the rail transportation use was reduced since the former is more versatile and it does not have a fixed minimum volume or a fixed route.

[38] These differences are kept although the rail tariff has increased due to the raise in the petroleum cost since 2002.

[39] With the shift of the agricultural frontier to the North East and North West of Argentina where distances to the port of Rosario are more than 350-400 km the transport cost incidence due to the rail use would decrease making those regions more competitive .

These river port terminals deliver 76% of the cereal, oilseeds protein meal and vegetable oil exports and the rest of the river port terminals deliver 4% of global cargo.

The grain solid by products storage volume in the Paraná Hydroway ports grew from 1.5 million tons to almost 8 million. The cargo boats rhythm, in turn, grew from 23 million tons per hour to 54 million tons between 1990 and 2007. This operative capacity growth also allowed a growth of the Up-River exports from 13.5 million tons to 54 million tons as regards grains and pellets, at no extra costs of storage or delays of the cargo boats stays (SAGPYA, 2007).

6. Future context for the Argentine soybean chain

At an international level, it is considered that soybean prices will keep growing. This conclusion is drawn from the soybean world market indicators (supply, demand, stock, consumption) and ratified by the wider expansion of the world demand regarding the supply. If finantial speculation is added, the positive trend towards rising prices gets stronger. This conclusion is drawn after an analysis of the following factors (USDA, 2010):

- Grain production growth motivated specially by the larger demand of developing countries (DC) (consumption growth of food and forage for animals due to the per capita income growth and diet changes).
- Global growth more accelerated in DC (city planning processes and middle class expansion).
- Oil consumption growth in China (34% imports growth during 2011 whereas in India there will be 21% imports drop).
- Bio-fuel demand growth in the developing countries and in the DC.

It is expected that the good crop perspectives determine that the area planted remains and/ or increases, depending onthe economic perspectives of other summer crops, specially corn.

In Argentina, it is expected that in the short–medium term the discontinuous purchase of soybean oil will continue to be done by China due to the commercial barriers implemented by the government of Argentina and China, except that an agreement between both countries is reached. This has caused that the oil exports go to other countries which buy this oil at a lower price[40]. So, the Argentine oil industry -which is one of the most efficient in the world– gets lower prices and, furthermore, works under its milling capacity.

At a national level and at the primary production link, there have been and there are conflicts between the soybean protagonists (soybean producers, agrochemical companies, commercial companies; etc) and different groups that question this "soybean production system", in Argentina[41]. Among those who question this "soybean production system", it can be mention the small and medium farmers, aborigine communities and countryside workers expelled by the spreading of soybean sowing. At the same time large company

---

[40] The argentine soybean oil quotation suffered a discount with respect to other markets like the Brazilian one of up to U$D 75.5 per ton (average of U$D 50 per ton). These figures affect the almost 6 million soybean oil tons that Argentina would export in the current season

[41] In the mid 70s and with the use of new varieties, the sowing of two annual crops (doble –sowing) was possible. These crops are combined with the wheat production (short cycle wheat) and therefore the double crop wheat-soybean appears. This process originated a larger agricultural process (cattle is left aside) but soybean also leaves aside crops like corn and sorghum

groups[42] are consolidated and thus finance great scale farming and livestock directed by contractors who rent the fields.

Also among those that question the advance of soybean production there are some professionals and environmentalists who show the farming effects as regards the distribution of the productive and environmental resources. Some of these concerns are:

- Growing loss of food sovereignty due to the soybean growing substitution.
- The soybean monoculture generates among other consequences: productive variability fall, low genetic variability, sanitary risks (new plagues, resistence to plagues and diseases).
- Soybean takes the soil nutrients, wich demands crop rotation, and in certain soils its sowing is not advaisable. Each year, soybean takes one million tons of nytrogen and more than 227,000 tons of phosphorus. Likewise, the organic material balance is negative since the mineralized carbone cannot be compensated by the one of the soybean remains.
- The soybean monoculture produce an intense soil degradation in the long run, with a losses of between 19 and 30 soil tons per hectare depending on management , soil sloping and weather .
- The no-till cropping implies strong applications of agrochemicals. Massive use of glyphosate (160 million liters in 2004/2005) and other agrochemicals (between 20 and 25 million liters of 2-4-D (herbicide), and nearly six million liters of atrazine and 6 million liters of endosulfan; these last two, insecticides.
- The product in wich glyphosate is the active principle (Round up) has a series of coadjutant that increase its toxicity considerably, specially the polioxietil amina (surfactant), whose acute toxicity is 3 to 5 times higher than the one of the glyphosate.
- The advance of the soybean over the native forests (massive deforestation and loss of biodiversity). Between 1998 and 2002, more than 2 million hectares of native forest were deforestated[43].

Thus, there is a scenery of progressive conflict, where soybean excellent prices are signals for producers to continue with the expansion of soybean, and for other social groups (small farmers, environmentalists, professionals, etc.) to press because of the environmental and social consequences of this expansion.

7. Conclusions

Beyond the positive aspects related to the primary production (more security than other crops specially due to weather conditions, considerable finantial rentability, etc) the soybean growth has had positive effectsin Argentina, like the strong investments in new technologies (no-till cropping) biotechnology (transgenic soybean, fertilizers and herbicides), intensive use of agricultural machinery, storage (silo bags), logistics (private ports), net organiation (of services and financing), industrial machinery development, biofuel and agro food development the significant income of currency for the economy, without forgetting to mention the levying of export duties.

---

[42] "Los Grobo" Group claims that they grow 150 thousand hectares in Argentina and attemps to control 750 thousand hectares in all the Southern Cone (Paraguay, Brazil and Uruguay).

[43] Beginning of the century: 100 million hectares in forests, at the end of XX century: 33 million hectares

Without minimazing the positive effects or the conflicts brought about by the soybean expansion in Argentina, this crop has also generated a strong dependence[44] -a scarce commercial diversification –since only one product (soybean bean and its derived products) represent 25% of the total exported by Argentina (14.041 million dollars) (INDEC, 2009). This exporting dependence gets worse because the diary products, fruit, vegetable oil and meat are considered "Low Technological Level", whereas milling products are labeled as "Middle-Low Technological Level" (the technological content is based on the expenses in I+D) (CEP, 2007). At a Gross Product Level, the soybean chain (2009) represents 5.4% of the national added value (0.77% in the two-year period 73/74).

The strong soybean dependence is also seen in indicators like the soybean crop area and the soybean oil and meal production. At a primary level, soybean covers 54% of the cultivated area (more than 18 million has) over the total implanted (28 million has). For the oil and meal production numbers are more significant: the soybean oil production in Argentina participates in the 79% of all oil manufactured and in the 93% of all the meal produce.

This shows that the oil argentine industry depends basically on soybean. It must be pointed out that the oil industry milling capacity is 57 million tons; 37-38 million of which were used in the year 2010.

This idle capacity cannot be covered with the imports since the current imports regulations mekes it too costly. Thus, the industry competitiveness will be limited to the crops primary production and to the fundamental characteristics of domestic low consumption of byproducts which allow strong exportable balance. This last one, shortened with the current domestic soybean use for biodiesel production.

Finally, the National State role must be examined and also the implemented policies in relation to the agricultural production. During 2011, due to higher expenses[45], the National State do not show any sign to eliminate[46] the exporting duties[47], especially the ones which belong to the soybean complex. These obligations represent (2010) 8.7% of the total taxes. This shows the importance of the exports duties and the dependence of these assets for the expense policies. The exporting duties were 7,400 million dollars; of which 6300 million dollars were collected by the soybean complex (2011).

8. References

Ablin, E., Paz, S. (2003). El mercado mundial de la soja, la República Argentina y los organismos genéticamente modificados, In: *Las Negociaciones comerciales multilaterales de la Ronda Doha. Desafíos para la Argentina.* Centro de Economía Internacional. (7-54). Instituto del Servicio Exterior de la Nación. Ministerio de Relaciones Exteriores de Argentina. Buenos Aires. Argentina

---

[44] The risks of non-tariff imposition barriers for genetically modified soybean must be considered.

[45] During 2009 and 2010 the State Budget result was negative (income lower than total expenses).

[46] In the National Legislature (Commission) a project to modify the exporting duties obligations is being debated.

[47] In the year 2002 export reimbursements were implemented: the soybean oil 1.6for the raw oil in bulk; 5%for refined oil in containers of 5kg and 0,7% for the raw oil. All the above mentioned refunds were left aside (0%) since November, 2005. The raw oil but the exported volumes of refined oil represent only 3% of all the soybean oils.

ACSOJA. (2009). Importancia económica de la soja. 15/08/2010. Available from: http://www.acsoja.org.ar/ampliarII.asp?intIdSection=88&intIdContenido=217.año

Bisang, R. (2003). Apertura Económica, innovación y estructura productiva: la aplicación de biotecnología en la producción agrícola pampeana argentina. In: *Desarrollo Económico*. Vol. 43 N° 171. (413-442). Buenos Aires. ISSN 0046-001X

Bisang, R., Sztulwark, S. (2007). Tramas productivas de alta tecnología y ocupación. El caso de la soja transgénica en la Argentina. In: Estructura productiva y empleo. Un enfoque transversal. Primera edición. (181-224). ISBN 978-84-36571-68-6 (web, pdf)

Brieva, S. (2006). *Dinámica socio-técnica de la producción agrícola en países periféricos: configuración y reconfiguración tecnológica en la producción de semillas de trigo y soja en Argentina, desde 1970 a la actualidad*. Tesis doctoral. Facultad Latinoamericana de Ciencias Sociales (FLACSO). (pp.292)

Cámara Industrial de Aceites de la Republica Argentina (CIARA). (2011). *Estadísticas de soja*. 10/02/2011. Available from: http://www.ciaracec.com.ar/estadísticas

Casini, C. (2002). Guía para almacenar grano en bolsas plásticas. In: *Información técnica Proyecto eficiencia de cosecha y postcosecha de grano*. (1-4). Instituto Nacional de Tecnología Agropecuaria (INTA) Manfredi, Córdoba.

Cámara Argentina de Biocombustibles (CARBIO). (2011). *Estadísticas de biocombustibles*. 13/02/2011. Available from: www.carbio.com.ar/es/?com=bio_estadísticas

Centro de Estudios para la Producción (CEP). 2007. "*Contenido tecnológico de las exportaciones argentinas 1996-2007. Tendencias de upgrading intersectorial*". 1/08/2010. Available from: www.cep.gov.ar/descargas_new/contenido_tecnológico_exportaciones_19962207.pdf

Civitaresi, H., Granato, M. (2003). El complejo oleaginoso argentino: algunos factores determinantes de su desempeño exportador. *Revista Argentina de Economía Agraria*. Vol. VI, 1, (Otoño 2003), pp. 37-47), ISSN 0327-3318

Dohlman, E., R. Schnepf., Bolling, C. (2001). Agriculture in Brazil and Argentina: Developments and Prospects for Major Field Crops. In: *Washington, DC: US Department of Agriculture, Economic Research Service. Agriculture and Trade Report*. WRS-0-1-3. 5/02/2011. Available http://www.ers.usda.gov/publications/wrs013/wrs013g.pdf.

EUMERCOPOL (2007). EU-Mercosur. Agrifood Systems Comparison. In: Proyecto. Documento de Trabajo Work Package 4. Buenos Aires.

Facultad de Agronomía, Universidad de Buenos Aires (FAUBA). (2004). *Informe Final LART/FAUBA. Patrones espaciales y temporales de la expansión de soja en Argentina*. Universidad Nacional de Buenos Aires. Buenos Aires, Argentina.

FAO-SAGPYA. (2004). *Análisis de las cadenas de maíz y soja en Argentina, con vistas a exportación de mercaderías OVM y no OVM en el marco del art. 18.2 a) del Protocolo de Cartagena sobre seguridad de la biotecnología*. Proyecto FAO-SAGPYA. TCA/ARG 2903. Documento N° 2. Dirección Nacional de Mercados Agroalimentarios. Organización de Naciones Unidas para la Agricultura y la Alimentación. Buenos Aires, Argentina.

Fundación Producir Conservando. Capacidad de almacenamiento. (2010). Available from: http:/www.producirconservando.org.ar

Giancola, S., Salvador, L., Covacevich, M., Iturrioz, G. (2010). Análisis de la cadena de soja en la Argentina. En: *Estudios Socioeconómicos de los Sistemas Agroalimentarios y Agroindustriales N° 3*. INTA, Buenos Aires. ISSN 1852-4605.

Ghida Daza, C. (2002). Alternativas para el almacenaje. Silo bolsa de grano seco. In: *Maíz, actualización 2002*. Información para extensionistas. INTA Marcos Juárez.

Gutman, G., Lavarello, P. (2003). La trama de oleaginosas en Argentina. In: *Estudios sobre el sector Agroalimentario. Estudio 1.EG.33.7 Componente B: Redes Agroalimentarias, Tramas B-3*. Bisang R., y G. Gutman. CEPAL-ONU, Buenos Aires.

Instituto Nacional de Estadística y Censos (INDEC). (2009). Estadísticas macroeconómicas. 20/02/2011. Available from: http://www.indec.mecon.gov.ar

Instituto Nacional de Tecnología Agropecuaria (INTA). Consejo del Centro Regional Santa Fe., (2009). Documento institucional. El avance de la soja en la Argentina y la sostenibilidad de los sistemas agrícolas. In: *INTA Centro Regional Santa Fe*. Febrero 2011. http://www.inta.gov.ar/reconquista/crsantafe/docsoja.htm

Lopez, G., (2008). *Vamos al grano?. El Rol del Estado en el comercio granario argentino* (primera edición). Editorial Cúspide. ISBN 9789870588023. (288 pp).

Nardi, M., Davis, T., (2006). Supply chain management in Argentina and Brazil and the Effect on World Agricultural Competitiveness. *Presentation given to trading companies*. Buenos Aires, March 2006. 92 pp.

Obschatko, E., (1997). *Articulación productiva a partir de los recursos naturales. El caso del complejo oleaginoso argentino* (Primera edición). Documento de trabajo N° 74. CEPAL, Buenos Aires. UNDP Project RLA 88/039. Buenos Aires, Argentina.

Oficina Nacional de Control Comercial Agropecuario (ONCCA). 2006. Existencia física de granos al 31 de Diciembre 2006. 6/02/2011. Available from: http://www.onnca.gov.ar

Quaim, M., Traxler, G. (2002). Roundup Ready Soybeans in Argentina: Farm Level, Environmental and Welfare Effects. *Proceedings of Conference ICABR on Agricultural Biotechnologies: New Avenues for Production, Consumption and Technology Transfer*. Ravello, Italia. Julio 2002.

SAGPyA (2007). Informe de fletes y preliminar de transporte de granos. 12/02/2011. Available from: http://www.minagri.gob.ar/publicaciones.php

Schavarzer, J., Tavosnanska, A. (2007). *El complejo sojero argentino. Evolución y perspectivas*. Centro de Estudios y Perspectivas de la Argentina (CESPA). Universidad de Buenos Aires. Facultad de Ciencias Económicas. Documento de trabajo N° 10. Buenos Aires.

Tavarez, C. (2004). Factores críticos a Competitividade da Soja no Parana e no Mato Grosso. Companhia Nacional de Abastecimiento. Brasilia. 9 pp.

Trigo, E., Cap, E. (2006). *Diez años de cultivos genéticamente modificados en la Agricultura Argentina*. Consejo Argentino para la Información y el Desarrollo de la Biotecnología- ARGENBIO, Buenos Aires, Argentina. 53 pp

United State Department of Agriculture (USDA). (2010). USDA data base: commodities and trade. In: USDA Foreign Agricultural Service. 29/01/2011. Available from: www.fas.usda.gov/commodities.asp and /trade.asp

6

Mineral Nutrition

Vlado Kovacevic[1], Aleksandra Sudaric[2] and Manda Antunovic[1]
[1]University J. J. Strossmayer in Osijek, Faculty of Agriculture
[2]Agricultural Institute Osijek
Croatia

1. Introduction

Sixteen nutrient elements are essential for the growth and reproduction of plants. The source of carbon (C) and oxygen (O) is air, while water is source of hydrogen (H). Ninety-four percent or more of dry plant tissue is made up of C, H and O. Remaining thirteen elements, represent less than 6 percent of dry matter, are often divided in three groups (Johnson, 1987). The primary nutrients are nitrogen (N), phosphorus (P), and potassium (K). Secondary nutrients are sulfur (S), calcium (Ca) and magnesium (Mg). Micronutrients are required by the plant in very small amounts. They are iron (Fe), manganese (Mn), boron (B), zinc (Zn), copper (Cu), molybdenum (Mo) and chlorine (Cl). Nutrient removal of soybean by tone of grain and correspondingly biomass are about 100 kg N, from 23 to 27 kg P_2O_5, from 50 to 60 kg K_2O, from 13 to 15 kg CaO, from 13 to 16 kg MgO and considerable lower amounts of the other nutrients. In general, the fertilizer requirements for soybean are typically less than for other crops such as maize and wheat. Bergmann (1992) reported adequate concentrations of nutrients in dry matter of fully developed leaves at the top plant without petioles at the end of blossom as follows: 4.50-5.0 % N, 0.35-0.60 % P, 2.5-3.70 % K, 0.60-1.50 % Ca, 0.30-0.70 % Mg, 25-60 ppm Zn, 30-100 ppm Mn, 25-60 ppm B, 10-20 ppm Cu and 0.5-1.0 % Mo.

2. Nitrogen

Symbiotic nitrogen (N) fixation had important role in supplying of leguminose plants including soybean, by N. It is estimated that by this source is possible to bind from 40 to 300 kg N/ha/year (Bethlenfalvay et al., 1990). Field studies by Bezdicek et al (1978) showed that soybeans are capable of fixing over 300 kg N/ha when the soil is low in available N and effective strains of *Bradyrhizobia* are supplied in high number. Also, part of leguminose needs for N is settled by its uptake in mineral forms, mainly in NO_3^- and NH_4^+ forms. Soybean contains in mean from 1.5 to 1.6 % and from 6.5 to 7.0 % N in dry matter of aboveground part and grain, respectively (Hrustic et al., 1998). N amounts removal from soil by soybean depending on numerous external and internal factors. For forming of 1 t of grain and correspondingly vegetative mass of soybean is needed about 100 kg N.
Worldwide some 40 to 60 million metric tons (Mt) of N_2 are fixed by agriculturally important legumes annually, with another 3 to 5 million Mt fixed by legumes in natural ecosystems, providing nearly half of all the nitrogen used in agriculture (Hungria & Campo, 2004). Therefore, biological dinitrogen fixation by leguminous plants is a significant source

of available nitrogen in both natural and managed ecosystems (Galloway et al., 1995) that contributes to soil fertility and replaces the use of synthetic nitrogen fertilizer. The host plant provides carbon substrate as a source of energy, and bacteria reduce atmospheric N_2 to NH_3 which is exported to plant tissues for eventual protein synthesis (Vincent, 1980). Nitrogen fixation occurs in different intensities in soil, during which the energy of plant assimilates is used, and because of this, bacterial activity forms unbreakable relationship with plants. The proportion of nitrogen derived from fixation varies substantially from zero to as high as 97%, and most estimates fall between 25% to 75% (Deibert et al., 1979; Keyser & Li, 1992, Russelle & Birr, 2004).

N is mobile in plants and it is quickly translocate from old to young organs. For this reason, symptoms of N deficiency (first lightgreen and later greenyellow colours of leaves) obtain on the older leaves. In the more over stages it is found falling off the flowers and pods (Vrataric & Sudaric, 2008). Excess of N had unfavorable impacts on soybean productivity, mainly due to susceptibility to diseases, low temperatures and drought. Symptoms of N oversupplies are increasing of height of plants, longer internodies and lodging incidences. Soybean is the most susceptible leguminose to nitrate oversupplies. Under these conditions inhibition of nodule forming and nitrogenase activities were found Harper & Gipson, (1984). Also, high nitrate in apoplast of soybean had effect on pH increasing, immobilization of iron and developing of iron chlorosis in soybean (Hrustic et al., 1998).

N supplies of soybean could be estimated by number and activities of bacterial nodules of genus *Rhisobium* and *Bradyrhisobium*, contents of total and mineral N in oil, nitratreductase activities etc. Inadequate N supplies are possible to correct by mineral fertilization. Activities of bacteria are reduced under good supplies of soil by N (as results either high N fertilization or favorable conditions for organic matter mineralization) and acid soil pH.

Recommendations for soybean fertilization are depended on soil test results and planned yields. Under conditions of the northern Croatia N recommended quantities are mainly in range from 60 to 90 kg N/ha mainly in spring. Using of N as urea in autumn over 100 kg/ha resulted by absence of nodule bacteria or minimizing their amounts (Vrataric & Sudaric, 2008). By testing 12 localities in fertile soils (chernozem and similar soil types) of Vojvodina (Serbia) was found that inoculation had considerable more impacts on yields of soybean compared to N fertilization and that using 90 kg N/ha was not found nodule on soybean root (Belic et al., 1987; Relic 1988 – cit Vrataric & Sudaric, 2008). Based on experiences from very fertile soils in Ohio (Johnson, 1987), soybean is not recommend for N fertilization in case of sufficient amounts of N-fixing bacteria and only in first growing of soybean on individual soil recommendation is applying 45 kg N/ha. Also, in Illinois mineral N fertilization had not effects on soybean yields even in cases of band fertilization close to soybean rows. Also, N fertilization was superfluous for maize in soybean-maize rotation (Welch et al., 1973). However, the experiences from USA are not possible to applying in less fertile soils of middle and eastern Europe.

Soil acidity is often limiting factor of the symbiotic nitrogen fixation process. Soils with low pH values lack calcium, and have surplus of toxic aluminium, so that soybean roots in acidic soils don't have mucous coating on surface which purpose is to dissolve root pectines, enables root hair curling and root hair penetration by bacteria. This is very important during the first few days after inoculation that is after sowing inoculated seed. Therefore, soils with pH value less than 5.5 (acidic soils) are not suitable for soybean growing, because they lack necessary conditions for development of useful bacteria whose growth is slowed down or completely enabled. Strains found on soybean roots in this type of soils are mostly ineffective, and when cut in half are green in colour. Situation is completely opposite in fertile neutral or mildly alkaline soils like chernozem. In these types of soil nitrogen fixing bacteria have not only good conditions for development, but also they can survive in large numbers for many years after soybean was grown. In such soils it is not necessary to perform seed bacterisation if soybean is in rotation every four years.

In case of low effects of inoculation on nodule bacteria development it is recommend top-dressing with 50 kg N/ha in form of calcium ammonium nitrate (27% N) in term close to flowering or at beginning of flowering (Vrataric & Sudaric, 2008).

Organic manures cannot alone meet the heavy demands of nutrients in intensive soybean production because of their limited availability and restricted nutrient supply. A complementary use of organic manures and mineral fertilizers may meet the goal of adequate and balanced supply of required nutrients to crops. The soybean grain yield with recommended NPK fertilization and 25 kg N/ha + 1 t neem cake/ha combinations was significantly more than the other only chemical and organic source of nutrition (Table 1).

Soybean grain quality (crude protein and oil contents) and grain yield as affected by fertilization (in. = inoculated; neem cake = n.c.; FYM = farm-yard manure 5 t/ha; NPK 20:60:40 = recommended NPK-fertilization)							
Treatments	Percent		t/ha	Treatments	Percent		t/ha
(a-e)	Protein	Oil	Yield	(f-j)	Protein	Oil	Yield
a) Control (in.)	34.42	19.03	0.69	f) b + 1 t n.c.	37.92	18.85	1.52
b) 25 kg N/ha	37.92	16.00	0.93	g) c+ 1 t n.c.	37.04	19.65	1.23
c) 50 kg N/ha	37.04	18.72	0.89	h) b + 5 t FYM	38.06	16.83	1.57
d) 5 t FYM/ha	35.73	19.23	0.80	i) c+ 5 t FYM	37.63	18.17	1.20
e) 1 t n. c/ha	35.44	17.03	1.07	j) NPK 20:60:40	38.94	18.92	1.33
LSD 5 % (a-j)	ns	1.78	0.27	LSD 5 % (a-j)	ns	1.78	0.27

Table 1. Effects of inorganic and organic sources of nutrients on grain quality and yield of soybean (Saxena et al., 2001)

3. Phosphorus

Phosphorus (P) contents in plants are in wide range, mainly from 0.1 to 0.8 % P in dry matter. Reproductive organs, especially of leguminose plants contain high levels of P about 0.6 % P. Uptake of P into plants is intensive in the early stages of growth and in period forming of generative organs (Hrustic et al., 1998). Store of P in plants, especially in grain, are mainly in form of fitine acid. P efficiency is in close connection with water and temperature regimes in soil. Under optimal soil moisture P uptake can be up to three-fold higher than in dry soil. Also, oversupplies of water, cold weather and low pH reducing P uptake in plants.

P removal by plants is mainly from 10 to 45 kg P, while by soybean is from 15 to 30 kg P/ha/year. The end of growth is the first symptom of P deficiency. Leaves are dark green and in the later stage develops chlorosis and violet color as result of increasing antociane synthesis. Necrotic spots, drying and falling of the leaves is the latest stage of P deficiency. Active nodules (dark pink center) of N-fixing bacteria are absent or few in number under conditions of P deficiencies. Also, decreasing of protein and chlorophyll synthesis was found.

Excess of P is rare. Plants reducing growth and dark frowning spots in leaves were observed. Intensity of plant development increasing and as results are the earlier flowering, grain forming and senescence. Oversupplies of P could be reason for some nutritional unbalances, for example Zn, Fe, Mn, Cu and B deficiencies.

P, mainly in combination with N and K as NPK fertilizers, can be applied broadcast and incorporated into the soil before sowing or applied as starter at sowing time. With low soil test P levels band application of fertilizer is more efficient than broadcasting. If applied as a starter, the recommend placement of the fertilizer is in band 2 inches to the side and 2 inches bellow the seed (Dahnke et al., 1992; Barbagelata et al., 2002). P materials such as triple superphosphate or from liquid or dry formulations of ammoniated phosphates are available to improve soil P status. However, organic soybean growing has restriction in P use and it is limited on rock phosphate or manures as sources of P.

Anetor and Akinrinde (2006) found that P deficiency in soil is an important growth-limiting factor in acidic alfisolof Western, Nigeria. Lime application may not be feasible for poor resourced farmers. However, the complementary benefits (liming and nutrient supply) of organic fertilizers and rock phosphates could sufficiently ameliorate acid soil conditions and greatly reduce P fertilizer cost for effective and sustainable soil fertility management.

Win et al. (2010) tested the P effects on three soybean cultivars (CKB1, SJ5 and CM60) based on the seed oil content (SOC) and the seed protein content (SPC) and to assess the physiological responses associated with changes in shoot P-utilization efficiency (SPUE). The experiment was carried out during 2008 and 2009 with a split-plot design at the Agronomy Department, Kasetsart University, Bangkok, Thailand. The main plots were for tested three P levels in a nutrient solution (0.5, 1.0 and 2.0 mM P), with subplots for the three soybean cultivars. The results indicated that at maturity, the P levels of 2.0 mM P decreased SPUE by 27% compared to that of 0.5 mM P (the control). SOC was not significantly affected by the P level. Relative to the control, the P nutrition levels of 1.0 and 2.0 mM P significantly decreased SPC by 4% and 5%, respectively. There were no significant differences in SOC between varieties. The SPC of CKB1 was 8% greater than that of SJ5 but showed no significant difference to that of CM60 (Table 2).

Zheng et al., (2010) reported effectiveness of P application in improving regional soybean yields under drought stress of the 2007 growing season in Northern China including Heilongijang, Jilin and Liaoing Provinces. Total soybean acreage of this region was around 4.5 million ha, which accounts for about 5% of the total soybean acreage in the world (FAOSTAT, 2009). Contemporary climate change is characterized by increase in frequency and intensity of drought. Total 118 soybean fields throughout Hailun County of Northern China. Regression trees analysis showed that regional soybean yield variability was mainly induced by soil available phosphorus and the amount of P applied, which explained 16.3 and 15.2% of the yield variation, respectively. The productivity of soybean over the region did not increase when P application rate reached a threshold of 55.67 kg/ha (Zheng et al., 2010).

Effects of P levels in nutrient solution and cultivars on soybean status (DM = dry matter; Sh = shoot; Prot. = protein)										
R5 stage (Σ = Total)			Maturity stage				Maturity stage			
Dry matter (g/plant)			Dry weight (g/plant)		P-utilization* (Q=quotient)		Phosphorus (mg/plant)		Seed %	
Sh.	Root	Σ	Σ	Shoot	Qa	Eff. b	Sh.	Seed	Oil	Prot.
Effects of P levels (m*M* P) in nutrient solution										
0.5	14.1c	1.35b	15.4c	25.8b	24.6b	0.213a	5.21a	118b	47.9b	16.9	40.2a
1.0	20.5a	2.36a	22.9a	36.0a	34.4a	0.139b	4.77a	258a	72.3a	18.1	38.5b
2.0	18.2b	1.96a	20.2b	33.3a	31.6a	0.117b	3.79b	292a	64.4a	17.5	38.3b
Effects of soybean cultivars										
CKB1	22.2a	2.54a	24.7a	34.2	32.5	0.156	4.89a	236	59.7a	16.9	40.4a
SJ5	14.2c	1.42b	15.7c	27.1	25.9	0.147	3.72b	212	52.5b	17.3	37.1b
CM60	16.4b	1.74b	18.1b	33.8	32.2	0.167	5.17a	221	72.4a	18.3	39.4a
Duncan's multiple range test (within column, means by the same letter are not significantly at 5 % level): CV %										
a	12	13	12	21	21	12	10	13	14	17	2
b	8	17	8	21	22	14	12	16	15	7	2

* a Shoot P-utilization quotient = plant shoot dry weight/mg P in plant shoot of P
  b Shoot P-utilization efficiency (eff.) = [(shoot DM)^2/shoot P content]

Table 2. Effects of P on seed oil and protein contents and P use efficiency in three soybean cultivars (Win et al., 2010)

4. Potassium

Potassium (K) is essential nutrient for plant growth. K concentrations in dry matter of plants vary between 1.0 and 6.0 % and more and are generally higher than those of all other cations. The exact function of K in plant growth has not been clearly defined. By numerous investigations were found that K stimulates early growth, increases protein production, improves the efficiency of water (drought resistance), improves resistance to diseases, insects and stalk lodging (Kovacevic & Vukadinovic, 1992; Rehm & Schmitt, 1997).

Soils mainly containing enormous amounts of K, but depending on soil types, 90-98 percent of total K is unavailable. Slowly unavailable K is thought to be trapped between layers of clay minerals (Johnston, 1987; Rehm & Schmitt, 1997). K deficiency is encountered mostly on light, usually acid soils with a low cation exchange capacity or on soils with a high content of three-layered clay minerals often loess soils with illite clay (Bergmann, 1992).

Soybean requires large amounts of K and K deficiencies are easy to recognize (edge necrosis of leaves – the margins of leaflets turn light green to yellow) and correcting them is inexpensive as K is to lowest-cost major nutrient. K deficiencies as result of strong K fixation and high levels of available magnesium (Mg) were found on heavy hydromorphic soils of Sava valley area in Croatia. By ameliorative KCl fertilization yields of maize and soybean drastically increased due to improved plant nutritional status (Vukadinovic et al., 1988; Kovacevic & Vukadinovic, 1992; Kovacevic 1993; Kovacevic & Grgic, 1995, Kovacevic & Basic, 1997).

The K deficiency in soybeans was found on the drained gleyols which had inadequate rates of the exchangeable K and Mg (low K and high Mg status). These soil characteristics affected correspondingly K and Mg status in soybean plants (Tables 3 & 4).

Soybean (the upermost full-developed threfoliate leaf before anthesis) and soil status (means of four fields)								
The state farm (year)	Soybean (K-deficieny symptoms)				Grain yield	Soil status (0-30 cm of depth)		
	Leaf status (precent in dry matter)					pH		mg/100 g (AL-method)
	P	K	Ca	Mg	(kg/ha)	H₂O	KCl	P₂O₅	K₂O
Zupanja (1988)	0.35	0.98	1.20	0.73	1930	7.33	6.91	6.6	15.9
Vinkovci (1989)	0.57	1.05	2.22	2.14	780	7.75	6.87	28.0	12.2
Jasinje (1988)	0.38	0.87	1.30	1.11	1410	7.68	7.20	10.6	10.2
N. Gradiska (1990)	0.32	1.16	1.82	0.92	2060	7.76	6.91	7.9	16.6

Table 3. Plant and soil status (drained gleysol): symptoms of K deficiency in soybean (Kovacevic et al., 1991)

Soybean (the upermost full-developed threfoliate leaf before anthesis) nutritional status								
The state farm and date of sampling	Percent in dry matter				mg/kg (ppm) in dry matter			
	P	K	Ca	Mg	Zn	Mn	Fe	Al
	Chlorotic soybeans (K-deficiency symptoms): means of two samples/field							
Zupanja (June 19, 1987)	0.45	2.02	1.74	2.91	13.0	28.0	609	588
Vinkovci (June 13, 1986)	0.73	0.66	1.16	1.65	19.0	25.0	600	
Jasinje (June 13, 1986)	0.52	0.70	1.11	1.29	27.0	200.0	220	
Mean	0.57	1.13	1.34	1.95	19.7	84.3	476	
	Normal soybeans (oasis in the chlorotic soybeans): means of two samples/field							
Zupanja (June 19, 1987)	0.25	2.87	1.69	1.78	12.0	28.0	386	309
Vinkovci (June 13, 1986)	0.59	1.13	1.48	1.25	17.0	26.0	260	
Jasinje (June 13, 1986)	0.59	1.06	1.25	1.11	18.0	248.0	180	
Mean	0.48	1.69	1.47	1.38	15.7	100.7	275	

Table 4. Nutritional status of normal and chlorotic (K-deficiency symptoms) soybeans (Kovacevic et al., 1991)

Response of soybeans to ameliorative KCl- fertilization (the upermost full-developed threfoliate leaf before anthesis)												
Fertilization (spring 1986)			The 1986 growing season					The 1987 growing season				
N	P₂O₅	K₂O	Yield (t/ha)	Leaf (% in dry matter)				Yield (t/ha)	Leaf (% in dry matter)			
				P	K	Ca	Mg		P	K	Ca	Mg
0	0	0	2.43	0.37	0.72	1.84	1.62	1.45	0.35	0.91	1.69	1.35
120	120	180	2.40	0.36	0.87	1.87	1.49	1.48	0.32	0.99	1.37	1.35
120	120	990	2.83	0.35	1.28	1.76	0.74	1.88	0.32	1.29	0.92	0.77
LSD 5%			0.27	ns	0.22	ns	0.25	0.29	ns	0.16	0.30	0.31

Table 5. Response of soybeans to potassium fertilization (Katusic et al. 1988; cit. Kovacevic & Basic, 1997)

KCl in spring 1987	Fertilization (KCl) impacts on soybean: grain yield and leaf (the upermost full-developed threfoliate leaf before anthesis) K and Mg (on dry matter basis) status– the growing seasons 1987-1989								
	1987			1988			1989		
K_2O	Yield	Leaf (%)		Yield	Leaf (%)		Yield	Leaf (%)	
kg/ha	kg/ha	K	Mg	kg/ha	K	Mg	kg/ha	K	Mg
150	1280	0.57	1.60	1800	0.82	1.18	780	0.60	2.16
1000	2700	1.90	0.95	2350	1.74	0.84	1470	0.75	1.79
2670	2550	2.28	0.78	2740	2.22	0.52	2530	1.17	1.41
LSD 5%	270	0.20	0.20	450	0.09	0.18	240	0.07	0.21
LSD 1%	360	0.27	0.27	600	0.13	0.24	320	0.09	0.27

Table 6. Response of soybean plants to potassium fertilization (Kovacevic & Vukadinovic 1992)

Katusic et al., (1988; cit. Kovacevic & Basic, 1997) applied increasing rates of KCl on Cerna drained gleysol. Soybean responded by yield increases for 16 % and 30 %, for the first and the second year testing, respectively. Soybean under unfertilized and usual fertilization was contained in mean 0.82 % K (acute K-deficiency with correspondingly symptoms) and 1.49 % Mg. Soybean nutritional status was considerable improved by K fertilization (mean 1.29 % K and 0.76 % Mg) – Table 5.

Kovacevic & Vukadinovic (1992) tested response of soybean and maize to increasing rates of potassium application in KCl form on silty clay gleysol developed on calcareous loess. Low levels of exchangeable K, high levels of exchangeable Ca and Mg and strong K fixation were found by the soil test (Vukadinovic et al., 1988; Kovacevic & Vukadinovic, 1992). Also, clay fraction (35.2 % of soil) composition was as follows: vermiculite/chlorite 30 %, smectite 30 %, mixed layer minerals 20 %, illite 15 % and kaolinite 5 % (Richter et al., 1990). By ameliorative K fertilization soybean yields were increased drastically (3-y means: 1286 and 2607 kg/ha, for the control and the highest rate of K) and they were in close connection with improvement of leaf K and Mg status (Table 6 and Fig. 1).

Fig. 1. Soybean status (middle of July 1989) on the control (left) and the highest rate of K (2670 kg K_2O/ha in spring 1987) application (right) – the data in Table 6 (photo V. Kovacevic)

In Ontario, Canada, studies looked at the response of soybeans to potassium fertilizer as related to K leaf tissue levels (Reid and Bohner, 2007). The data collected during that study formed the basis for updated critical and normal values for potassium in soybeans. Below a leaf K concentration of 2.0% (on dry matter basis), most of the plots showed a response to added K fertilizer. Above this level, most of the plots were unresponsive. Based on the results of these experiments and other similar studies, the critical concentration for K in soybean tissue was established at 2.0% and the maximum normal concentration from 2.5 to 3.0%. According this criterion, in our investigations under strong K-fixing conditions (Table 5) only by application of enormous K rates leaf-K concentrations were increased to normal level. However, in spite of considerable improvement of soil and plant K status, yields of high-yielding soybean cultivar were less than 3.0 t/ha (Table 5).

Long-term studies conducted on integrated nutrient management in soybean-wheat system (Singh & Swarup, 2000) revealed that continuous use of FYM along with recommended NPK for 27 crop cycle not only restricted K mining by reducing non-exchangeable K contribution to grain formation but also enchanced K uptake to the system (Table 7).

Fertilizer K added in 27 crop rotation, total K uptake, available and non-exchangeable K status in soil (maize crop was discontinued in the system since 1995)						
Treatment	Potassium (kg K /ha)				Contribution of	
	K added	Available K status		Total K	non-echangeable K	
	in 27 cycles	Before 1971	After 1999	uptake	kg K/ha	%
a) Control	0	370	252	3247	3129	96.4
b) 100 % N	0	370	263	4418	4311	97.6
c) 100 % NP	0	370	235	10067	9932	98.7
d) 100 % NPK	2117	370	308	11826	9647	81.6
e) d + 5 t FYM/ha	4142	370	324	14094	9906	70.3

Table 7. Removal and addition of K during 27 crops cycle of soybean-wheat-maize (fodder) cropping system (Singh and Swarup, 2000)

Morshed et al. (2009) applied six treatment of potassium (unfertilized, 50%, 70%, 100 % , 125% and 150% of recommend rate based on soil test) on equal N, P and S fertilization in Dhaka (Bangladesh) during Rabi season 2004-2005. By application of the highest K rate grain yield of soybean was increased for 83%. Slaton et al., (2009) found close connection of soybean response to K fertilization (five rates from 0 to 148 kg K/ha) and Mehlich-3-extractable soil K in eastern Arkansas. Experiments were established on silt loams at 34 site-years planted with a Maturity Group IV or V cultivar. Mehlich-3-extractable soil K ranged from 46 to 167 mg K/kg and produced relative soybean yields of 59 to 100% when no K was applied. Eleven sites had Mehlich-3-extractable K < 91 mg K/kg and all responded positively to K fertilization. Soybean grown in soil having 91 to 130 mg K/ g responded positively at nine of 15 sites. Mehlich-3 soil K explained 76 to 79% of the variability in relative yields and had critical concentrations of 108 to 114 mg K/kg, depending on the model. Based on these investigations, Mehlich-3-extractable K is an excellent predictor of soil K availability for soybean grown on silt loams in eastern Arkansas.

Gill et al. (2008) reported that imbalance and inadequate nutrient supply particularly devoid of K is main reason for low productivity and quality of soybean in India.

Yin and Vyn (2004) conducted field experiments at three locations in Ontario, Canada from 1998 through 2000 to estimate the critical leaf K concentrations for conservation-till soybean on K-stratified soils with low to very high soil-test K levels and a 5- to 7-yr history of no-till management. For maximum seed yield, the critical leaf K concentration at the initial flowering stage (R1) of development was 2.43 %. This concentration is greater than the traditional critical leaf K values for soybean that are being used in Ontario and in many U.S. Corn Belt states.

Nelson et al. (2005) compared response of soybean to foliar-applied K fertilizer and preplant application. Potassium fertilizer (K_2SO_4) was either broadcast-applied at 140, 280, and 560 kg K/ha as a preplant application or foliar-applied at 9, 18, and 36 kg K/ha at the V4, R1-R2, and R3-R4 stages of soybean development. Soybean grain yield increased 727 to 834 kg/ha when K was foliar-applied at 36 kg/ha at the V4 and R1-R2 stage of development in 2001 and 2002. Foliar-applied K at the R3-R4 stage of development increased grain yield but not as much as V4 or R1-R2 application timings. Foliar K did not substitute for preplant K in this research. However, foliar K may be a supplemental option when climatic and soil conditions reduce nutrient uptake from the soil.

Numerous studies investigated fertilization effects on soybean grain yield, but few focused on oil and protein concentrations. Haq & Mallarino (2005) determined fertilization effects on soybean grain oil and protein concentrations in 112 field trials conducted in Iowa from 1994 to 2001. Forty-two trials evaluated foliar fertilization (N-P-K mixtures with or without S, B, Fe, and Zn) at V5-V8 growth stages. Seventy trials evaluated preplant broadcast and banded P or K fertilization (35 P trials and 35 K trials). Replicated, complete block designs were used. Foliar and soil P or K fertilization increased ($P < 0.05$) yield in 20 trials. Foliar fertilization increased oil concentration in one trial and protein in one trial but decreased protein in two trials. Phosphorus fertilization increased oil concentration in two trials and protein in five trials but decreased oil in five trials and protein in two trials. Potassium fertilization increased oil in four trials and protein in two trials but decreased oil in two trials and protein in two trials. Total oil and protein production responses to fertilization tended to follow yield responses. Fertilization increased oil production in 20 trials and protein production in 13 trials. Fertilization that increases soybean yield has infrequent, inconsistent, and small effects on oil and protein concentrations but often increases total oil and protein production.

Potassium is known to play an important role in protecting the plants against drought stress. Quantity and distribution of rainfall in the major soybean regions in India is responsible for yield fluctuations about plus/minus 20% among years in comparison with national average yield of 1 t/ha. For example, K fertilization in level of 112 kg K_2O/ha resulted by soybean yield increases for 0. 2 t/ha in normal year (1980) and for 1.2 t/ha under drought stress conditions (1981). Profit from K fertilization was 44 and 259 USD/ha, for 1980 and 1981, respectively (Johnson, 1984). For this reason, K fertilization can help in curtailing the yield loss on account of drought.

There are several materials available to supply K to the soil and potassium chloride is the most economical form. However, certified organic soybean production is limited to the use of potassium sulfate or manures to supply K.

5. Secondary nutrients

Calcium, magnesium and sulfur comprise the secondary nutrient group. Documented deficiencies of these three elements are few (Council for Agricultural Science and Technology, 2009).

5.1 Calcium

Plant species differ greatly in their Ca needs. Total Ca contents in plants are mainly in range from 0.5 to 1.0% in dry matter. The Ca uptake of plants influenced by Ca status and pH value of the soil and by the concentrations of other cations, especially K and Mg. Lack of Ca in legumes prevents the development of the nodule bacteria, thus affecting N fixation. Ca containing materials are using in correction of soil pH from acid to close to neutral. Soil pH between 5.5 and 7.0 is optimal for symbiotic N fixation in soybean root nodules by *Bradyrhizobium japonicum* bacteria. Under these soil pH availability of nutrients such as N and P and microbial breakdown of crop residues are favorable. Calcium deficiency is unlikely if soil pH is maintained above 5.5 (Council for Agricultural Science and Technology, 2009).

5.2 Magnesium

The total Mg content in plants is generally between 0.1 and 0.5 % in dry matter. Mg is the central atom of chlorophyll and it is vital for photosynthesis, biological production and conversion of matter in the plant metabolism. Mg deficiency occurs on strongly leached diluvial sandy acid soils with a low cation exchange capacity. Mg deficiency can be induced not only by low Mg status but also by high concentrations of other cations, for example H^+, K^+, NH_4^+, Ca^+ and Mn_2^+(Bergmann, 1992). In Croatia were found nutritional problems of K uptake by soybean and maize induced by oversupplies of Mg and strong K-fixing (Kovacevic and Vukadinovic, 1992). Vrataric et al. (2006) reported increases of soybean yield for 5 %, contents of grain protein for 0.7% and oil for 0.7% due to foliar application of 0.5 % $MgSO_4$ (Epsom salt) solution on eutric cambisol. Importance of Mg in yield increases of field crops in Europe reviewed by Uebel (1999).

Vrataric et al. (2006) tested response of six soybean cultivars (*Kuna, Una, Nada, Ika, Lika and Tisa*) to foliar fertilization (FF) with Epsom salt ($MgSO_4.7H_2O$; 5% w/v solution in amount 400 L/ha) on Osijek eutric cambisol. The fertilization was applied on standard fertilization either once or two times (treatment designations FF 1x and FF 2x, respectively), while untreated plots were as a control (standard fertilization). The first FF was made in the soybean stage V2-V3 and the second FF ten days later before the R1 stage of soybean. The amounts of added nutrients were as follows (kg/ha): 3.2 MgO and 2.3 kg S, as well as 6.4 MgO and 4.6 kg S, for the treatment FF 1x, and FF 2x, respectively. In the growing season 1999 was by 22% higher compared to 1998. Yield of *Ika* cultivar was by 23% higher compared to *Una*. FF resulted by moderate yield increases up to 5% compared to the control. Differences of yield between FF 1x and FF 2x were non-significant. Oil contents were higher in the 1998 and 2000 (mean 21.27%) compared to 1999 and 2001 (mean 20.55%), while differences among cultivars (from 20.77% to 20.96%) were non-significant. In general, FF resulted by moderate but significant oil content increases (20.45%, 21.15% and 21.12%, for the treatment 0, FF 1x and FF 2x, respectively). Protein contents were significantly different among years from 38.53% (2000) to 39.38% (2001) and among the cultivars from 38.30%

(Lika) to 39.48% (Nada). ESFF resulted by significant increases of protein contents (38.62%, 39.11 and 39.21% for 0, FF 1x and FF 2x, respectively). Impacts of the fertilization on soybean yields were shown in the Table 8.

Cul-tivar (B)	Foliar fertilization (factor A) with Epsom salt (F1x in V2-V3 and F2x = F1x + ten days later) effects on grain properties (four year means: 1998-2001) of soybean cultivars (factor B)											
	Yield (t/ha)			X B	Oil content (%)			X B	Protein cont. (%)			X B
	0	F1x	F2x		0	F1x	F2x		0	F1x	F2x	
Kuna	3.70	3.98	3.99	*3.89*	20.5	21.2	21.1	*20.9*	38.0	38.7	38.9	*38.5*
Una	3.52	3.64	3.66	*3.61*	20.4	21.3	21.3	*20.9*	39.1	39.3	39.4	*39.2*
Nada	3.77	4.01	4.04	*3.94*	20.4	21.2	21.2	*21.0*	39.1	39.6	39.7	*39.5*
Ika	4.44	4.45	4.45	*4.45*	20.3	21.0	21.0	*20.8*	38.4	39.1	39.2	*38.9*
Lika	3.55	3.97	3.65	*3.63*	20.7	21.0	21.0	*20.9*	38.0	38.5	38.5	*38.3*
Tisa	3.89	4.05	4.11	*4.02*	20.4	21.2	21.2	*20.9*	39.2	39.6	39.6	*39.5*
X (A)	3.81	3.97	3.98		20.5	21.2	*21.1*		*38.6*	*39.1*	*39.2*	
LSD 5%	A: 0.13	B: 0.10	AB: 0.19		A: 0.36	B: ns	AB: 0.46		A: 027	B: 0.31	AB: 0.46	
LSD 1%	0.35	0.13	0.31		0.84		0.66		0.50	0.48	0.66	

Table 8. Impacts of foliar fertilization with Epsom salt ($MgSO_4.7H_2O$; 5% w/v solution in amount 400 L/ha) on soybean properties - four year means (Vrataric et al., 2006)

5.3 Sulphur

Soils of humid and semi humid areas mainly contain total sulphur (S) in range from 100 to 1000 mg/kg, a range that is similar to that of total P. It is divided in inorganic and organic forms but in most soils organically bund S provides the major S reservoir. S in organic matter can be divided into two fractions, carbon bonded S and non carbon bonded S. The inorganic form of S in oil consists mainly of sulphate. In arid regions soils may accumulate high amounts of salts such as $CaSO_4$, $MgSO_4$ and $NaSO_4$. Sulphate like phosphate is adsorbed to sesquioxides and clay minerals, although the binding strength for sulphate is not a strong as that for phosphate. Under waterlogged conditions, inorganic S occurs in reduced forms such as FeS, FeS_2 and H_2S. Oxidation of S results int he formation of H_2SO_4 and is promoting factor of additional soil acidification. Sulphate acid soils are mainly extremely low pH and very rich in exchangeable Al. Soil acidification by addition of elemental S is recommend for depressing the pH of alkaline soils (Mengel & Kirkbi, 2001).
Sulphur contents in plants are mainly in range from 0.1 to 0.5 % in dry matter. S uptake by plants is in sulphate form, but plants can absorb S also in gaseous form as SO_2. Sulphate must first be reduced by the plant to sulfide before it can be incorporated mainly into S-containing amino acids methionine and cistine. S deficiencies in plants are relatively rare because of the constant inputs of sulphate with NPK fertilizers and presence of SO_2 in precipitation (acid rain). Soybeans use a considerable amount of sulfur. S deficiency is mainly occurs during cool, wet weather on highly leacheable sandy soils that are low in organic matter and in little industrialization areas. In some cases are possible damages due to S excess caused by acid rain (Bergmann, 1992).
Sarker et al., (2002) tested effects of fertilization of soybean by S and B alone or in combination up to 50 kg S/ha and u to 4.0 kg B/ha. Yield, protein and oil contents of soybean grain where significant when S and B were applied individually but their

interaction were not significant. The highest biological yield and most of the yield atributes were obtained for the treatment combination of 30 kg S/ha and 1.0 kg B/ha.

6. Micronutrients (Zn, Mn, Fe, Cu, B, Cl, Mo)

6.1 Zinc

Soybean, maize and flax are the most susceptible field crops to Zn deficiency. It is often found on sandy soils low in organic matter, on high soil pH and calcaric soils, as well as on soils rich in available P. Cold and wet weather promoting Zn deficiency. N improving, while Fe and especially P, decreasing Zn uptake by plants. The first symptom of Zn deficiency in soybean is usually light green color developing between the veins on the older leaves. New young leaves will be abnormally small. Bronzing of the older leaves may occur. When the deficiency is severe, leaves may develop necrotic spots. Shortened internodes will give plants a stunted, rosetted appearance (Dahnke et al., 1992).

Zinc is essential element in metabolism of protein, carbohydrate and lipids. Zinc is compound of some enzymes (carboanhydrase, glutamat and malat-hydrogenase, alcalic phosphatase, proteinase, peptidase, etc.). Zinc has influences on auxine synthesis, intensity of respiration and uptake of Cu, Mn and especially P. Also Zn contributing to increase resistance to viruses diseases, drought and low temperature stress. Soil and leaf testing use in diagnosis of Zn status in plants. Also, important is P/Zn ratio.

Incorporation of anorganic Zn in form of $ZnSO_4.7H_2O$ (2-22 kg Zn/ha) or organic Zn in chelate form (03-6.0 kg Zn/ha), as well as foliar fertilization (0.5% solution of zinc sulfate) could be use for corrections of Zn deficiencies.

Nutritional disorders were found in soybeans grown on Osijek calcareous eutrical cambisol. Growth retardation and chlorosis were accompanied with the alkaline or a neutral soil reaction. By the foliar diagnosis zinc deficiency was found. Zinc deficiency was promoted by the excess of phosphorus or iron/aluminum in plants while K deficiency was accompanied with the excess of magnesium uptake. For example, chlorotic soybean contained in means only 16 ppm Z in dry matter (into normal soybean 27 ppm Zn). At the same time, the P:Zn ratio was 239 (the normal levels are under 180), while Fe:Zn ratio was 34 (the normal levels are under 15). The analogous values for the normal soybeans were 150 and 7, respectively (Kovacevic et al., 1991). The higher soil pH and oversupplies of plant available P are factors promoting Zn deficiency in soybean (Table 9).

Soybean: The uppermost full-developed trifoliate leaf (June 6, 1990)								Soil (0-30 cm of depth); mg/100 g = AL-method			
Percent in dry matter				mg/kg in dry matter				pH		mg/100 g	
P	K	Ca	Mg	Zn	Mn	Fe	Al	H_2O	KCl	P_2O_5	K_2O
Chlorotic and growth-retarded soybean (means of three samples)											
0.39	2.36	2.51	0.88	16.3	124	547	301	7.47	6.60	62.6	45.3
Normal soybean (oasis at the same plot: means of five samples)											
0.37	2.52	1.93	0.68	26.8	86	195	147	6.70	5.90	42.5	54.5

Table 9. Plant and soil status (eutric cambisol of Agricultural Institute Osijek): symptoms of Zn deficiency in soybean (Kovacevic et al., 1991)

Response of soybean to fertilization: pods/plant (P/P), grain/pod (G/P), 100-grain weight and grain yield												
Treatments 1-6 (kg/ha)				P/P	G/P	100gw	Yield	Treatments 7-12 (kg/ha)				P/P	G/P	100gw	Yield
N	P2O5	K2O	Zn			g	t/ha	N	P2O5	K2O	Zn			g	t/ha
0	0	0	0	31.6	1.96	11.5	1.46	60	80	30	0	51.4	2.03	14.1	1.89
30	0	0	0	34.7	1.98	11.6	1.64	90	40	30	0	51.9	2.12	14.4	1.90
30	40	0	0	45.1	1.99	12.6	1.74	90	60	30	0	65.9	2.14	16.6	2.23
30	40	30	0	43.7	1.99	12.6	1.79	90	80	30	0	47.2	2.07	13.5	1.77
60	40	30	0	45.1	2.01	12.7	1.76	90	80	60	0	45.0	1.84	12.3	1.71
60	60	30	0	44.5	2.05	12.9	1.82	90	80	60	25	68.9	2.14	16.1	2.48
LSD (1-12) 5 %				8.51	ns	2.29	0.34	LSD (1-12) 5 %				8.51	ns	2.29	0.34

Table 10. Effect of N, P; K and Zn application on yield attributes and grain yield of soybean (Singh et al., 2001)

Rose et al. (1981) were studied response of four soybean varieties (*Lee, Forrest, Bragg* and *Dodds*) to foliar zinc fertilization ($ZnSO_4.7H_2O$ before flowering) at three sites in central and north-west New South Wales. At Narrabri one spray of 4 kg/ha gave a yield increase of 13 %. At Trangie and Breeza, two spray each of 4 kg/ha increased yield by 57 % and 208 %, respectively. Lee was the least responsive variety at each site and Dodds and Forrest the most responsive to applied zinc. Zinc fertilizer increased plant height, leaf-Zn, oil contents (at two sites) but decreased leaf-P. Leaf-P in untreated plots was indicative of varietal sensitivity to zinc deficiency both within and between sites.

Singh et al., (2001) tested twelve nutrient combinations comprising of three levels each of nitrogen (30, 60 and 90 kg N/ha), phosphorus (40, 60 and 80 kg P_2O_5/ha) two levels of potassium (30 and 60 kg K_2O/ha) and a single level of zinc (25 kg Zn/ha) along with control. Zinc fertilization in combination with N, P and K significantly increased the growth attributes and grain yield of soybean, The highest number of pods per plant and grain yield were obtained with the joint application of N, P, K and Zn at the rates of 90, 80, 60, and 25 kg/ha, respectively (Table 10).

6.2 Iron

Leguminose plants have higher needs for Fe in comparison to cereals. Fe participating in numerous metabolic processes including protein synthesis. Under F deficiency conditions were found high levels of low-molecular N substances, especially amino acid arginine. Soybean is susceptible to Fe deficiency. Fe deficiency is a common yield limiting factor for soybean grown on high-pH, calcareous soils, as well as on some seasonally poorly drained soils. Cool and wet periods are promoting Fe deficiency. Iron may be unavailable for root absorption, not transported after absorption, or may not be utilized by the plant.

In Iowa and Minnesota, over ten million dollars in potential soybean production were lost annually due to iron chlorosis (Fleming et al., 1984). With the potential increase in alkalinity of Texas soils due to irrigation, reduced soybean production may become a problem. The problem could result from decreased yield per acre or from acreage with decreased productivity due to increased alkalinity. Iron deficiency is not easy or inexpensive to correct in the field. According to Gray et al. (1982) it would take five tons of sulfuric acid per acre to neutralize one per cent calcium carbonate in a 16.5 cm layer of soil.

Fe deficiency results in a characteristic interveinal chlorosis in new leaves and can cause substantial yield loss in soybean. In some years, developed during early growth stages and disappears as the plants mature. In more severe cases, chlorosis can persist throughout the entire season. There is wide variation in susceptibility to Fe deficiencies among soybean varieties.

Soybean in chlorotic areas had lower leaf chlorophyll concentrations, stunted growth, and poor nodule development relative to nonchlorotic plants. Also, compared to nonchlorotic areas, soil in chlorotic areas had greater soil moisture contents and concentrations of soluble salts and carbonates (Hansen et al., 2003).

Correcting Fe chlorosis often requires a combination of management practices including variety selection, application of Fe fertilizers with the seed (for example iron chelate Fe-EDDHA) or foliar treatment with 1 % solution of ferrous sulfate.

Franzen and Richardson (2000) tested soil factors affecting iron chlorosis of soybean. Total 12 sites of Red River valley of North Dakota and Minnesota were studied in the 1996-1998 period. Calcium carbonate equivalence and soluble salts were most often correlated with chlorosis symptoms.

Plant response to iron chlorosis varies between cultivars and environmental conditions (Coulombe et al., 1984; Gray et al., 1982). The reduction of iron at the root surface from Fe to Fe is an adaptive mechanism which iron efficient plants use to overcome iron deficiency. Soybean cultivars like Hawkeye have been shown to be rather effective in facilitating iron uptake by this method (Brown & Jones, 1976). Iron uptake is (1) as iron in association with chelate molecules and (2) as ionic iron after chelate splitting. Iron efficient plants have a much increased rate of iron uptake after chelate splitting during iron deficiency chlorosis (IDC)-induced stress; iron inefficient plants do not (Romheld & Marschner, 1981). Iron efficient and iron inefficient plants reportedly are distinguishable in terms of extent of iron uptake as a function of phosphorus content in the soil. Chaney & Coulombe (1982) reported that increased phosphorus inhibited the increase in iron uptake of inefficient types and slightly reduced iron uptake of efficient types.

Goos & Johnson (2003) found considerable differences of resistance of soybean varieties to iron clorosis. (Table 11) Growing of more tolerant varieties is solution for alleviation of nutritional problems induced by iron deficiency.

Soybean varieties characterizing low chlorosis score (CS <2.3)			Soybean varieties characterizing high score (CS >3.2)		
Variety	Originator	CS	Variety	Originator	CS
Trail	N.D. AES	1.7	IA 2042	Iowa AES	3.7
Danatto	N.D. AES	2.0	IA 2041	Iowa AES	3.7
MN 0201	Minn. AES	2.0	MN 1103SP	Minn. AES	3.7
92 M10	Pioneer	2.0	MN 101SP	Minn. AES	3.7
IA 1005	Iowa AES	2.0	Minnatto	Minn. AES	3.5
Jim	N.D. AES	2.2	IA 2050	Iowa AES	3.3
MN 0203 SP	Minn. AES	2.2	IA 2033	Iowa AES	3.3
Mn 0302	Minn. AES	2.2	MN 2101SP	Minn. AES	3.3
Nornatto	N.D. AES	2.2	IA 2050	Iowa AES	3.3
MK 0649	Richland Organics	2.2	Parker	Minn. AES	3.3
CV = 31.7; LSD 5% = 1.0					

Table 11. Chlorosis scores of soybean varieties in Minnesota 2003 (Goos & Johnson, 2003): score 1.0 = no chlorosis, 5 = most severe chlorosis (choice 20 extremely of 104 tested genotypes)

Silman and Motto (1990) tested under greenhouse conditions in nutrient solutions influences zinc on the growth and composition of an Fe-efficient (*Hawkeye*) and Fe-inefficient (*PI-54619-5-1*) soybean genotypes in various levels of Fe. In general, increased Zn levels resulted in growth reduction in both genotypes with the Fe-inefficeint plants being more sensitive to Zn level. The Fe-efficient genotype had a higher Fe content than the Fe-inefficient at corresponding treatment levels.

6.3 Manganese

Plants vary considerably in their Mn requirements and levels only 20 ppm often being sufficient for normal plant growth. Levels of Mn in plants vary between species and soil properties more than those of other nutrients. Plants generally absorb Mn from the soil as Mn_2^+. Mn is important in plant metabolism because of its redox properties and thus its ability to control oxidation, reduction and carboxilation reactions in the carbohydrate and protein metabolism.

Mn deficiency causes soybean plants to be stunted. The leaves are yellow to whitish but with green veins. Mn deficiency is most pronounced in cool weather on alkalic and slow alkalic soils rich in organic matter. Soil pH is the most important factor affecting Mn availability because it is extremely soluble at low pH and insoluble at high pH levels.

Foliar fertilization of young and moderately young crops with 8-15 kg $MnSO_4$/ha as 1-2 % solution (2-3 applications) is recommendation for prevention of Mn deficiency on soils with a high pH value.

Manganese deficiencies in oats and soybeans were reported by Willis (1928), primarily in spots in the coastal plain of North Carolina. This problem was associated with very high soil pH, and thus this observation was likely the first evidence of "overliming." The soils in the coastal plain are inherently low in manganese, especially the more poorly drained ones, since manganese can be reduced and leached in the soil-forming process (Cox, 1965). Interveinal chlorosis is a clear symptom of manganese deficiency. Cox (1968) used both extractable manganese by the Mehlich-1 extract and soil pH and developed a yield response prediction and manganese soil test interpretation for soybeans. Extractable concentrations at the critical level, which varied from 3 to 9 with Mehlich-1 depending on pH, were sufficient to be measured readily. Critical levels of manganese in soybean leaves at various growth stages and effective rates of fertilization for correcting manganese deficiency in soybeans reported by Mascagni & Cox (1985a, 1985b).

6.4 Copper

Copper (Cu) is seldom deficient in soil. Only on soils high in organic mater and under conditions pH above 6.0 would Cu likely be deficient. The color of legume and forage plants deficient in Cu tends to be grayish-green, blue-green or olive green. The internodes become shortened to produce a bushy type of plants (Sauchelli, 1969). Soybean has low requirements for Cu.

Williams (1930) noted that crops grown on muck soils in North Carolina often responded to copper application. This observation was researched in detail by Willis (1937), Willis & Piland (1936) and these researches centered on the aspect that copper may be a catalyst in oxidation-reduction processes in soils.

6.5 Boron

Total boron (B) contents in soils are into range of 20 to 200 mg/kg dry weight, most of which is unavailable for plants. The available, hot water soluble fraction in soils adequately supplied with B ranges from 0.5 to 2.0 mgB/L. Soluble B consists mainly of boric acid which under most soil pH conditions (ph 4-8) is undissociated. In soils of arid and semi arid regions, B may accumulate to toxic concentrations in the upper soil layer because lack of drainage and the reclamation of such soils requires about three times as much water as that of saline soils. Soil organic matter is closely associated with the accumulation and availability of B in soils (Mengel & Kirkby, 2001).

B deficiency leads to disturbance of growth and development of plants. B is known to influence carbohydrate metabolism, sugar transport, the nucleic acid and protein household, N metabolism, flower formation and pollen germination, water household, energetic processes of phosphorylation and dephosphorylation, etc. Conditions that favour B shortage are high pH (7.0-8.0), soils low in organic matter, drought, high concentrations of iron and aluminium hydroxide. Difference between adequate and toxic concentrations of B is very small. Soybean is very susceptible to B toxicity. Alfalfa and sugar beet have high requirements in B. Broadcasting and incorporation of 0.5 to 1.0 kg B/ha (for example, the most commonly used borax) is satisfied for needs of crops in rotation for a few years (Berrgmann, 1992; Mengel & Kirkby, 2001).

6.6 Chlorine

Chlorine (Cl) is in group of elements which can have a beneficial effects on plant growth. Plant tissues usually contain substantial amount of Cl often in range of 2 to 20 mg/kg dry weight. Soils considered low in Cl are below 2 mg water soluble Cl/kg soil which is rare. The effects of excess Cl in plants are more serious problem. Crops growing on salt affected soils often show symptoms of Cl toxicity. These include burning of leaf tips too margins, bronzing, premature yellowing and abscission of leaves. Plant species differ in their sensitivity to Cl. Some leguminous species are very prone to Cl toxicity and using of sulfate instead of chlorine fertilizers is recommend (Mengel and Kirkby, 2001).

6.7 Molybdenum

Molybdenum (Mo) is an essential plant nutrient. The concentrations of Mo may vary from less than 0.1 to more than 300 ppm. Roots contain a greater proportion of Mg than aboveground part or seed. Molybdenum is needed by the soybean and other leguminose plant itself and also by the nitrogen-fixing *Rhizobia* bacteria in the soil. In contrast to other micronutrients, Mo availability increases with soil pH. Seldom is there Mo deficiencies with soil pH above 6.0. Since the element is critical for nitrogen fixation, the pale green or yellow plants are identical to a nitrogen deficiency. In this case, leaves generally begin to yellow first on the lower leaves. Needless to say, symptoms usually do not occur on soils high enough in nitrogen to make up for lack of nodule fixation (Holshouser, 1997). Efficiency of symbiotic N_2 fixation can be limited by micronutrient deficiencies, especially of molybdenum. Soybean generally responds positively to fertilization with Mo in soils of low fertility and in fertile soils depleted of Mo due to long-term cropping.

Sodium or ammonium molybdate are mainly used for correction of Mo deficiency either as a solid to the soil or by spraying on the foliage or by treating the seed. The first step, however, is always to establish the proper soil pH. The micronutrient can be supplied by seed treatment, however toxicity of Mo sources to *Bradyrhizobium* strains applied to seed as inoculant has been observed, resulting in bacterial death and reductions in nodulation, N2 fixation and grain yield. Therefore, use of seeds enriched in Mo could be a viable alternative to exterior seed treatment. Campo et al., (2009) demonstrated the feasibility of producing

Mo-rich seeds of several soybean cultivars, by means of two foliar sprays of 400 g Mo/ha each, between the R3 and R5 stages, with a minimum interval of 10 days between sprays (Table 12). In most cases, Mo-rich soybean seeds did not require any further application of Mo-fertilizer (Campo et al., 2009).

Soybean variety	Foliar fertilization by Mo (R3 = beginning of pod formation; R5 = beginning of pod filling; 2x = two spraying at R3+R5 with ½ dose)									CV %
	0	400 g Mo/ha			800 g Mo/ha			1600 g Mo/ha			
		R3	R5	2x	R3	R5	2x	R3	R5	2x	
	Mean contents of molybdenum in seed of soybean (ug/g)										
Embrapa 48	3g	23e	17f	31d	36c	25e	43b	39c	39c	61a	8.8
BRS 133	4f	22e	18e	35c	33d	26d	50b	50b	39c	82a	18.7
BRS 156	4f	19e	20e	35d	33d	35d	56b	52b	43c	81a	13.2

Table 12. Impacts of foliar fertilization on Mo contents in soybean grain (Campo et al., 2009)

7. Harmful elements (Cd, Cr, Hg and Pb) and heavy metal toxicities

Heavy metals are the intrinsic component of the environment. It is usually accumulated due to unplanned municipal waste disposal, mining and use of extensive pesticides. Other agro-chemicals uses as chemical fertilizer are the significant cause of elevation in environment.

Shute et al. (2006) reported results of greenhouse study regarding Cd and Zn accumulation in soybean. The highest dose of Cd (100 mg/kg) reduced plant height and dry weight (down to 40 % and 34 % of control, respectively), while the analogical data for the highest dose of Zn (2000 mg/kg) were 55 % and 70 %, respectively. With both metals present, the plants were approximately the same size as those treated with cadmium only. When both metals were added to the soil, 80-100 % of the cadmium and 46-60 % of the zinc were bioavailable. Concentrations of both metals were highest in root tissue (10-fold higher for Cd and up to 2-fold higher for Zn). Although relatively little Cd was translocated to pods and seeds, the seeds of all plants (including those from control and zinc-treated plants) had concentrations of cadmium 3-4 times above the limit of 0.2 mg/kg set by the Codex Alimentary Commission. This was surprising given that Cd in the soil was only 1 mg /kg well below the maximum allowable amount for agricultural soil.

The heavy metal content of municipal and industrial sewage sludge and swine manure lagoon sludge are quite high in Cu and Zn and cause a buildup of the elements in the soil and for this reason have potential toxicity to the environment. (King, 1986; King & Hajjar 1990). Physiological effects of zinc toxicity in soybean elaborated Fontes (1992) and Fontes & Cox, 1995, 1998). Borkert & Cox (1999) evaluated the effects of high concentrations of both Zn and Cu on soybean status. Miner (1997) looked at soil factors affecting plant concentrations of these elements in sludge-amended soils. When concentrations of heavy metals are high, knowledge of their solubility becomes important.

As soybeans are one of the principle sources of dietary intake in the Japanese population, the Codex Committee on Food Additives and Contaminants has proposed an upper limit of 0.2 mg/kg for cadmium concentration in soybean grain with aim of protection dietary uptake of harmful quantities of Cd (Arao et al., 2003).

Arao et al. (2003) tested Cd uptake and distribution of Cd in 17 soybean varieties grown in pots (three soils: Mid-Cd Soil, High-Cd Soil, Low-Cd Soil) and under field conditions in unpolluted soil (low-Cd field). The sources of cadmium pollution were thought to be mine waste in the case of the Mid-Cd Soil, and refining plant waste in the High-Cd Soil. The seed cadmium concentration was lowest for the En-b0-1-2 soybean variety, and highest for *Harosoy*. The seed cadmium levels of *Tohoku 128*, a cross between *Enrei* and *Suzuyutaka*, were intermediate between those of the parents (Table 13). For four soil types, containing from 0.2 to 6.5 mg kg⁻¹ extractable cadmium, the ranking of soybean genotypes based on seed cadmium level was similar, indicating that there is a genetic factor involved in the varietal differences in cadmium concentration. The lower levels of cadmium found in the seeds of certain varieties of soybean could be result from the combination of lower initial uptake and retention of higher levels of cadmium in the roots, thus limiting its translocation to the shoot.

Different actions can be undertaken in order to reduce the absorption of Cd by plants. The addition of amendments such as calcium carbonate, zeolite, and manganese oxide can reduce Cd uptake in plants. With that regard, zeolite was more effective in suppressing Cd uptake by plants than calcium carbonate or manganese oxide (Chen et al., 2000; Putwattanaa et al., 2010). Also, organic amendment such as farmyard manure and compost which contains a high proportion of humified organic matter can decrease the bioavailability of Cd and other heavy metals in soil (Li et al. 2006, Pichtel & Bradway, 2008; Tordoff et al., 2000). Shamsi et al. (2010) tested effects of potassium supplementation on alleviation of Cd toxicity in hydroponics experiment. K supplementation at a rate of 380 mg/l in combination either with Cd addition (1 ug Cd) or without Cd. K supplementation alleviated the reduction of growth, photosynthesis and nutrient uptake in Cd-treated soybean plants. It was concluded that Cd toxicity could be alleviated through enhanced K nutrition in soybean.

Soybean cultivars show significant differences in seed cadmium concentrations, primarily because of genetic rather than environmental factors. One-six of the total soybean produced in Japan exceeded 0.2 mg Cd/kg, the international standard proposed by the Codex Alimentarius Commission. Further, the soybean crops had considerably higher Cd contents than other field crops MAFFJ (2002). Sugiyama & Noriharu (2009) investigated the seed Cd concentrations in four soybean cultivars (*Suzuyutaka, Hatayutaka, Enrei* and *Kantou 100*) in pot experiment on Cd-polluted soil. In *Suzuyutaka*, which had high Cd concentrations in the seeds, the concentrations of Cd distributed from the shoots to the leaves was 67% and that distributed from the shoots to the seeds was 13%. In *Kantou 100* which had low Cd concentrations into seeds, 57% Cd was distributed from the shoots to the leaves and 21% from the shoots to the seeds. These results suggest that cultivars that have a low capacity for Cd accumulation in the roots have a mechanism that prevents Cd accumulation into seeds by promoting its accumulation in the leaves (Sugiyama & Noriharu, 2009).

Chromium (Cr) is a nonessential and toxic element to plants. Chromium interferes with several metabolic processes, causing toxicity to plants as exhibited by reduced seed germination or early seedling development (Sharma et al., 1995), root growth and biomass, chlorosis, photosynthetic impairing and finally, plant death (Scoccianti et al., 2006). Normal range of Cr is from 10 to 50 mg/kg depending on the parental material (Pandey & Pandey, 2008). Researchers have demonstrated experiments with plants associated with high levels of Cr. Thus, 1-5 ppm Cr present in the available form in the soil solution, either as Cr (III) of Cr (VI), is the critical level for a number of plant species. Increased Cr (VI) concentration of 10-800 mg/l in culture medium led to the detection of inhibited growth parameters. There was a reduction in growth, dry weight and vigour index in four soybean genotypes of soybean at 5 -200 mgl-1 concentrations of chromium, according to control application (Ganesh et al., 2009).

Testing of 17 soybean cultivars				High-Cd soil (66 days after sowing)		
Soybean Cultivar (1-17)	Pot experiment		Field	Soybean cultivar	Cadmium	
	High-Cd soil	Mid-Cd soil	Low -Cd		ppm	ug/plant
	Seed Cd (ppm in dry matter)			Leaves		
En-b0-1-2	1.43 a	0.46a	0.08 a	En-b0-1-2	5.5a	67.6a
Tamahomare	2.52 abc	0.70abc	0.10 ab	Tohoku 128	12.2b	152.8b
En-b0-01	1.96 abc	0.82bcd	0.10 ab	Suzuyutaka	12.9b	86.9a
Goyoukuromame	1.99 abc	1.16ef	0.10 ab	LSD 5%	1.6	27.1
Hayagin	2.22 abc	0.91cde	0.11 ab	Stem		
Enrei	2.09 abc	0.89cde	0.11 ab	En-b0-1-2	4.3a	48.0a
En-b2-110	2.06 abc	0.91cde	0.11 ab	Tohoku 128	10.3b	130.1b
Dewamusume	5.24 d	1.05def	0.12 b	Suzuyutaka	20.3c	120.1b
Tachiyutaka	3.29 c	1.47g	0.12 b	LSD 5%	5.1	51.2
En-N0-2	4.94 d	1.91h	0.13 b	Pod		
Tachinagaha	2.88 bc	1.17f	0.13 bc	En-b0-1-2	2.1a	8.4a
Nattousyouryuu	2.90 bc	0.59ab	0.13 bc	Tohoku 128	5.5ab	8.3a
Getenshirazu 1	1.72 ab	0.78bcd	0.13 bc	Suzuyutaka	13.7b	8.6a
EN 1282	5.33 d	2.21i	0.16 c	LSD 5%	9.4	5.9
Tohoku 128	2.83 bc	0.97cde	0.22 d	Total		
Suzuyutaka	7.46 e	1.50g	0.31 e	En-b0-1-2		124.0a
Harosoy	12.68 f	2.68j	0.40 f	Tohoku 128		291.3b
Average 1 - 17	3.61	1.14	0.15	Suzuyutaka		215.6a
LSD 5 %	1.35	0.28	0.03	LSD 5%		83.8

Table 13. Seed Cd concentrations of soybean varieties grown in pots (choice of High-Cd soil and Mid-Cd soil) and under field conditions (Arao et al., 2003)

Mercury (Hg) and his compounds are among the strongest phytotoxic substances and are also extremely dangerous to human and animals. It is a constituent of many crop protection agents. Non-contaminated soils contain only 0.003 to 0.03 mg Hg/kg. Hg levels of about 0.04 mg/kg in dry matter can be considered normal in plants. Maximum tolerance limit of 0.05 mg/kg in fresh matter is proposed for foodstuffs. Mercury uptake in plants is very slight because it is strongly sorbet in the soil, mainly by complexation with organic matter. Apart from growth inhibition, the symptoms of Hg toxicity include chlorosis, necrotic lesions and death. These are mainly results of severe root damage and the consequent inhibition of nutrient and water uptake. Since little Hg is translocated out of the root, there is a little danger of its entering to the food chain through the soil. The mobility of Hg and its uptake by plants can be greatly reduced by liming (Bergmann, 1992).

Lead (Pb) is major chemical pollutant of the environment, and is highly toxic for man. The major source of Pb pollution arises from petrol combustion. This source accounts for about 80% of the total Pb in the atmosphere. Pb is toxic because it mimics many aspects of the metabolic behavior of Ca and inhibits many enzyme systems. There is evidence that Pb pollution can induce brain damage in man and aggressive behavior in animals. Pb toxicity interferes with Fe metabolism and the formation of haem. The total Pb concentrations of agricultural soils lie between 2 to 200 mg/kg soil. Pb contamination very clearly follows the motorway areas. Vegetation at the side of the road may have levels of 50 mg/kg dry matter but in distance of only 150 m away from the motorway the level is normally about 2 to 3 mg/kg. Contamination occurs only on the outer part of plant seed or leaves and stem, and high proportion can be removed by washing (Mengel and Kirkby, 2001).

8. Genetic aspects of mineral nutrition of soybean

Plant varieties of the same species differ in absorption and utilization of nutrients from the environment. Varietal differences in the uptake of individual nutrients can be used as basis for the testing of both commercial varieties and selection materials under unfavorable soil conditions. An adequate distribution of soybean varieties based on their tolerance or susceptibility to less favorable conditions could contribute to better utilization of their yield potential (Saric, 1981; Saric & Loughman, 1983). According to Epstein (1976) agricultural intervention in the process of nature has two corresponding strategies: selection and genetic manipulation of the organism and modification of the environment. Many crops in Brazil have their yield improved thanks to the selection and breeding, especially in the large savanna (cerrado) region of Central Brazil. Soybean cultivation in the low-latitude acidic soils of Brazilian Savanach has become a reality since 1970's. Great contribution for this success has been achievements in soil science and plant breeding. There are however, constraints for sustainable production like high-Al and low-Ca in the deep layers of the soil. Measures can be taken to reduce the negative effects of acidity on plant growth are liming and selection of more tolerant genotypes.

Kastori (1978) found that Ca uptake was higher in *Corsoy* than in *Stella* and *Wilkin*. Later research showed that K uptake was the highest at the variety *Corsoy* (Kastori et al., 1979). Keoght et al., (1977) found lower N uptake in the varieties *Hill*, *Lee* and *Bragg* than in *Hood*. Queiroz et al. (1980) tested residual effects of P fertilizers on the yield of three soybean varieties over four years. The variety *Bossier* increased grain yield by 320 kg/ha, *Parana* by 640 kg/ha, whereas variety *Vicoja* did not show any response to P fertilization. Saric and Krstic (1982) tested ten soybean varieties 30 days in N-deficient nutrient solution. The variety *Yoslie Kataya 2* showed the lowest and the variety *Traverse* the highest N contents.

Kovacevic and Krizmanic (1987) tested 12 soybean genotypes of maturity group I (*Corsoy* and *Hodgson* as standard varieties and remaining ten are experimental lines from the F8 generation) under calcareous soil conditions. Grain yield of soybean genotypes ranged from 2.1 4 to 3.11 t/ha. The highest yield of *Vuka* and the lowest yield of *Os 155/82* on this soil may be due to the lowest Ca status by the former and the highest Ca status by the latter genotype (Table 14).

Spehar (1995a, b; 1999) studied genetic differences in the accumulation of nutrients in leaves and seeds of tropical soybean cultivars from diallel crosses with the cultivars *IAC-9*, *IAC-2*, *UFV-1*, *IAC-5*, *IAC-8*, *Vx5-281*, *IAC-7*, *Biloxi* and *Cristalina* under high and low Al-stress. The diallel analysis indicated that an additive-dominance model could explain the genetic differences among those genotypes for nutrient accumulation in leaves and seeds. The diallel analysis, although not conclusive, indicated that the mechanisms of mineral element accumulation in the leaves are not fully associated to those of accumulation in the seeds of soybeans. The expression of these characters is, however, dependent on mineral plant-stress.

Yield (t/ha) and leaf composition of 12 soybean genotypes (the uppermost full-developed trifoliate leaf at beginning of flowering) on calcareous soil (pH in KCl 7.35; CaCO$_3$ 7.93 %)

Soybean genotype	t/ha	Leaf (% in dry matter)				Soybean genotype	t/ha	Leaf (% in dry matter)			
		N	K	Ca	Mg			N	K	Ca	Mg
Corsoy	3.10	5.20	2.05	0.91	0.55	*Os 8*	2.30	5.20	1.85	0.76	0.41
Hodgson	2.89	5.78	1.98	0.90	0.44	*Os 9*	3.00	4.82	2.18	0.85	0.36
Vuka	3.11	5.58	2.01	0.76	0.35	*Os 45*	2.41	5.07	1.99	0.82	0.35
Podunavka	2.74	5.73	2.06	0.98	0.51	*Os 89*	2.51	5.60	2.15	0.88	0.41
Sava	2.71	5.59	2.12	0.80	0.40	*Os155/82*	2.14	4.91	1.99	1.12	0.61
Os 5	2.82	5.48	2.15	0.95	0.39	*Os442/83*	2.45	5.46	2.24	0.86	0.50
LSD 5%	0.26	0.12	0.08	0.16	0.09	LSD 5%	0.26	0.12	0.08	0.16	0.09
LSD 1%	0.35	0.17	0.11	0.22	0.12	LSD 1%	0.35	0.17	0.11	0.22	0.12

Table 14. Yield and nutritional status of 12 soybean genotypes on Osijek calcareous soil (Kovacevic & Krizmanic, 1987)

Sudaric et al. (2008) reported about the effectiveness of biological nitrogen fixation in soybean linked to genotype for four growing season in eastern Croatia. Fields study involved eight cultivars in two treatments (control - without ionoculation and inoculation by *Bradyrhizobium japonicum*). The obtain results suggested on significant positive effect of rhizobial inoculation on both nitrogen fixation indicators and grain yield at all tested soybean cultivars (Table 15). Significant differences among tested cultivars in each measured trait indicate genetic diversity of tested material in both potential of biological nitrogen fixation and compatibility cultivar by *B. japonicum* strain, as well. Tested cultivars with the best potential for nitrogen fixation (OS-1-00, OS-3-0, OS-3-I) had the highest grain yield increasing (14.4%, 14.3% and 14.0%, respectively). These results indicate that the cultivars with the favorable performances of biological nitrogen fixation could be used as the parents for development new cultivars that are able to accomplish high grain yield with lower nitrogen level in soil.

Phosphorus is a major limiting factor for crop production of many tropical and subtropical soils. In Brazilian soils, high productivities of soybean are achieved by soil amendment techniques, using lime and fertilizers, supplying the nutrients required for best crop performance. The yield potential is an intrinsic factor and depends on plant germplasm characters that can be modified by selection and breeding (Furlani et al., 2002). Differences in grain yield among soybean cultivars under field conditions for P-, K- and N-efficiencies , were also reported by Raper and Barber (1970), De Mooy et al. (1973), Sabbe & Delong (1998), Sarawgi & Tripathi (1998), Hanumanthappa et al. (1998; 1999) and Ogburia et al. (1999).

Plant efficiency for phosphorus uptake and utilization may contribute to improve crop yield potential in situations of low P availability. Furlani et al., (2002) evaluated and classified twenty nine soybean cultivars in relation to the response to phosphorus (P) levels in nutrient solution. P uptake and use efficiency were estimated by the variables: shoot and root dry matter (DM) yield, P-concentrations and contents in plant parts and P-efficiency index (EI). The experiment was conducted in a greenhouse, during 1999, at Campinas, State of São Paulo, Brazil. The experimental design consisted of randomized complete blocks, arranged in split-plots, with three replications. The main plots were the P levels in the nutrient solution (64.5; 129; 258 and 516 mmol L^{-1}), and the subplots were the twenty-nine soybean cultivars, grouped according days to maturity. Multivariate analysis showed high correlation among the variables shoot-DM, total-DM and shoot P-concentration and P-efficiency index (EI). Cultivars were classified in efficient-responsive (ER)¾ *IAC-1*, *IAC-2*, *IAC-4*, *IAC-5*, *IAC-6*, *IAC-9*, *Sta. Rosa* and *UFV-1*; efficient-non-responsive (ENR) ¾ *IAC-7*, *IAC-11*, *IAC-15*, *S. Carlos* and *Cristalina*; inefficient-responsive (IR) ¾ *IAC-8*, *IAC-10*, *IAC-14*, *Bossier* and *Foscarin*; and inefficient-non-responsive (INR) ¾ *IAC-12*, *IAC-13*, *IAC-16*, *IAC-17*, *IAC-18*, *IAC-19*, *IAC-20*, *IAC-22*, *Paraná*, *IAS-5* and *BR-4*. The efficient-responsive soybean cultivars showed the highest values for shoot and total DM and EI, and the lowest shoot P-concentrations.

Soybean cultivar	Properties of uninoculated (- = control) and inoculated (+) soybeans							
	Nodule number/plant		Above-ground part of plant				Grain yield t/ha	
			Dry matter weight (g)		Nitrogen (% N)			
	-	+	-	+	-	+	-	+
OS-1-00	0	44.6	4.29	6.19	1.84	2.65	3.54	4.05
OS-2-00	0	37.8	3.41	4.91	1.69	2.44	3.39	3.85
OS-1-0	0	38.2	3.34	4.80	1.71	2.46	3.53	4.02
OS-2-0	0	35.8	3.21	4.62	1.67	2.40	3.36	3.81
OS-3-0	0	49.2	4.24	6.11	1.84	2.65	3.42	4.00
OS-1-I	0	31.9	2.64	3.80	1.46	2.11	3.57	4.03
OS-2-I	0	36.3	3.24	4.67	1.65	2.38	3.83	4.33
OS-3-I	0	40.2	4.17	6.00	1.82	2.62	3.71	4.23
LSD 5%	4.0		1.13		0.22		0.31	
LSD 1%	5.3		2.02		0.41		0.36	

Table 15. Mean values of nitrogen fixation indicators and grain yield of 8 soybean cultivars (2004-2007; Osijek, Croatia) (Sudaric et al., 2008)

Ojo et al. (2010) tested 55 soybean genotypes under acid soil conditions in area of Umudike, Nigeria for two growing seasons. Highly significant differences in genotypic effects were observed for all the traits (days to 50% flowering, plant height at maturity, number of pods/plant, 100-seed weight and grain yield). Eight acid tolerant varieties were found (*Conqvista*, *TGX 1896-3F*, *TGX 1897-17F*, *TGX 1866-7F*, *TGX 1805-31F*, *Milena*, *Doko* and *TGX 1844-18E*) with a higher grain yield of >1.80tons/ha compared to <1.45tons/ha in the previously recommended varieties (*TGX 1485- 1D* and *TGX 1440-1E*). The result also showed the potential of the EMBRAPA genotypes in upgrading the TGX varieties for higher productivity. The eight identified acid tolerant varieties could therefore be explored in the development of improved high yielding soybean genotypes for production on acid soils of Nigeria.

9. Agronomic management practice and nutritional status of soybean

9.1 Fertilization

In general, mainly fertilization of soybean with nitrogen, phosphorus and potassium are common agronomic practice. Additional using the other nutrients are more exception than rule. In some cases application of the higher P and K rates are needed for achieving of satisfied yields of soybean.

Phosphorus and potassium are limiting factor of field crops yield on some hydromorphic soils in Croatia (Kovacevic, 1993; Kovacevic et al., 2007; 2011; Rastija et al., 2006). By application of the ameliorative rates of NPK fertilizer up to 3748 kg/ha level soybean yields were increased up to 32 %. Protein contents in soybean grain were independent on the fertilization, while oil contents were increased up to 0.66% compared to the control (Rastija, et al., 2006). In the second experiment, P and K applied separately up to 1500 kg/ha either P_2O_5 or K_2O and in their combination (1000 + 1000 kg/ha). Yields of soybean were increased up to 21% (influences of P), 17% (influences of K) and 30% (PK influences). However, protein and oil contents in grain were independent on fertilization (Kovacevic et al., 2007).

Soybean is generally responsive to fertilization with inadequate nutrient supplies. For example, grain yields of soybeans were increased by 40% and 34% as affected by the K and P fertilization, respectively (Table 16). According to status of the uppermost full-developed trifoliate leaf (Jones, 1967, cit. Bergmann and Neubert, 1976; Bergman, 1992) the adequate P, and high Ca and Mg status as well as low K status was found in the soybean leaves when ordinary fertilization was applied (Table 16). However, nutritional status of soybean was considerably improved when affected by the ameliorative fertilization. Calcium uptake by soybean leaves was high and it was practically independent on the fertilization. Also, the K fertilization influenced the Mg status in soybean leaves: it was decreased in relative amount by about 30 % compared to ordinary fertilization. More favorable relationship between K and Mg was associated with K fertilization: 1.13 and 3.20 for ordinary fertilization and the highest rate of added K, respectively (Table 16).

Rastija et al. (2006) applied four rate of ameliorative PK-fertilization on acid soil. As affected by the fertilization grain yields of soybean were increased up to 32%. However, yield differences among three ameliorative treatments were non-significant. Protein contents in soybean grain were independent on the fertilization, while oil contents were increased up to 0.66% compared to the control (Table 17).

Fertilization (March 22, 1990) by P and K rates on equal (kg/ha: 90 N + 137 P_2O_5+ 132 K_2O) NPK fertilization and soybean propreties (the growing season 1990: the uppermost full-developed trifoliate leaf before anthesis)										
K_2O kg/ha	Yield t/ha	Leaf concentrations (% in dry matter)				P_2O_5 kg/ha	Yield t/ha	Leaf concentrations (% in dry matter)			
		P	K	Ca	Mg			P	K	Ca	Mg
132	2.13	0.32	1.17	1.80	1.04	195	2.86	0.33	1.24	1.86	1.05
433	2.69	0.32	1.49	1.82	0.92	325	2.71	0.35	1.39	1.89	1.04
735	2.98	0.32	1.65	1.77	0.83	585	2.57	0.38	1.52	1.94	0.98
1337	2.81	0.33	2.01	1.66	0.79	1105	2.52	0.49	1.80	1.92	0.91
2532	2.82	0.33	2.37	1.85	0.74	585	2.60	0.41	2.01	1.80	0.81
LSD 5%	0.49	0.01	0.13	0.19	0.08	LSD 5%	0.49	0.01	0.13	0.19	0.08
LSD 1%	0.66	0.02	0.17	0.26	0.11	LSD 1%	0.66	0.02	0.17	0.26	0.11

Table 16. Response of soybean to ameliorative P and K fertilization (Kovacevic, 1993)

Kovacevic et al. (2007) tested response of soybean to ameliorative P and K fertilization alone or in their combination. As affected by applied fertilization soybean yields were increased up to 21% (influences of P), 17% (influences of K) and 30% (PK influences). However, protein and oil contents in grain were independent on fertilization (Table 18).

Fertilization (April 23, 2004)			Soybean properties (the 2005 growing season)				
	kg/ha		(t/ha)	Percent in dry matter			
Treatment	P_2O_5	K_2O	Grain	Grain		Leaves*	
			yield	Protein	Oil	P	K
Control	125	82	3.88	41.92	20.33	0.530	2.67
PK-1	375	248	4.87	40.89	20.80	0.537	2.71
PK-2	625	414	4.73	41.42	20.62	0.571	2.85
PK-3	875	582	4.98	40.64	20.99	0.487	2.73
PK-4	1125	746	5.14	41.94	20.73	0.501	2.69
LSD 5%			0.72	n.s.	n.s.	n.s.	n.s.
			* the uppermost full-developed threefoliate leaf before anthesis				

Table 17. Residual impact of PK-fertilization on soybean properties (Rastija et al., 2006)

Kovacevic et al. (2011) reported residual impacts of increasing rates of PK-fertilization up to 1000 kg P_2O_5/ha and 672 K_2O/ha in spring 2004 and liming by granulated fertdolomite (24.0 % CaO + 16.0 % MgO + 3.0 % N + 2.5 % P_2O_5 + 3.0 % K_2O) in autumn 2007 on soybean status in the growing season 2010. As affected by liming yields of soybean were increased for 18 % (means 3279 and 3854 kg/ha, for unlimed and limed plots, respectively). Also, grain quality parameters were improved by liming (thousand grain weight were 151.8 and 168.3 g; protein contents were 35.24 and 39.06 %, respectively), while oil contents were decreased (23.84 and 22.62 %, respectively). However, impact of P and K fertilization was considerably lower in comparison with liming (Table 19).

Fertilization (April 23, 2004)*				Soybean properties (the 2005 growing season)				
		kg/ha		kg/ha	Percent in dry matter			
Treatment		P_2O_5	K_2O	Grain	Grain		Leaves*	
				yield	Protein	Oil	P	K
a	Control	125	82	3600	41.27	20.74	0.537	2.81
b	P-1	625	82	3580	40.76	20.83	0.513	2.86
c	P-2	1125	82	3460	41.57	20.64	0.593	3.15
d	P-3	1625	82	4360	41.74	20.52	0.603	2.88
e	K-1	125	582	4010	41.13	21.20	0.520	2.87
f	K-2	125	1082	4200	40.21	21.10	0.547	2.95
g	K-3	125	1582	4080	40.59	21.10	0.573	3.29
h	P2K2	1125	1082	4670	40.28	21.14	0.593	3.05
			LSD 5%	370	n.s.	n.s.	0.055	0.30
			LSD 1%	510			n.s.	n.s.
* for next year: 80 N + 125 P_2O_5 + 82 K_2O				* the uppermost full-developed trifoliate leaf before anthesis				

Table 18. Residual influences of NPK-fertilization on soybean properties (Kovacevic et al., 2007)

Residual effects of fertilization and liming on soybean (cultivar *Lucija*) grain yield and grain quality in the 2010 growing season										
Factor B*: Fertilization (April 2004)		Factor A**: Lime(t/ha) (Oct. 2007)		Mean B	Factor A**: Lime t/ha (Oct. 2007)		Mean B	Factor A**: Lime (t/ha) (Oct. 2007)		Mean B
kg/ha		0	10		0	10		0	10	
P_2O_5	K_2O	Grain yield (kg/ha)			Protein contents (%)			Oil contents (%)		
a	0	0	3141	3730	3422	36.12	37.68	37.68	23.70	22.49	23.10
b	250	168	3231	3837	3534	34.92	36.84	36.84	23.58	22.61	23.09
c	500	336	3285	4047	3666	35.73	37.59	37.59	23.58	22.29	22.94
d	750	504	3352	3826	3589	34.59	36.67	36.76	24.32	22.90	23.61
e	1000	672	3387	3860	3624	34.83	36.88	36.88	24.02	22.79	23.41
Mean A		3279	3854		35.24	37.15		23.84	22.62	
LSD 5% LSD 1%		A: 209 B: 156 AB: ns 482 ns			A: 0.63 B: ns AB: ns 1.46			A: 0.53 B: ns AB: ns 1.23		

* Ameliorative fertilization by NPK 10:30:20 (a-e) on ordinary fertilization; N added by NPK-fertilizer were equalized with CAN (calcium ammonium nitrate: 27% N);
** ferdolomite (24.0 % CaO + 16.0 % MgO + 3.0 % N + 2.5 % P_2O_5 + 3.0 % K_2O) 10 t/ha

Table 19. Residual effects of PK-fertilization (April 2004) and liming (Oct. 2007) on soybean (Kovacevic et al., 2011)

Response of soybean to phosphorus and potassium fertilization on yield and at the uppermost trifoliate leaves status								
Impacts of P (triple superphosphate) fertilization				Impacts of K (KCl form) fertilization				
P rate	Soybean yield (kg/ha)		Leaf P (%) at R2 stage		K rate	Soybean yield (kg/ha)		Leaf K (%) at R2 stage	
kg/ha	1997	1998	1997	1998	kg/ha	1997	1998	1997	1998
0	615	700	0.23	0.28	0	914	1180	1.51	2.27
5	814	890	0.26	0.35	9	973	1280	1.58	2.87
10	826	960	0.44	0.35	18	1092	1299	2.34	2.64
20	925	1338	0.44	0.57	36	1188	1320	2.36	2.54
30	1188	1667	0.46	0.51	54	1559	2236	2.66	2.64
40	1585	2433	0.46	0.40	72	2294	2725	2.62	2.62
50	2443	2731	0.44	0.35	90	2246	2773	2.71	2.70
60	2598	2814	0.50	0.35	108	2544	3164	2.25	2.65
70	2713	2938	0.50	0.38	135	2520	2815	2.21	2.61

P treatments received a blanket application of 108 kg K/ha; K treatments received a blanket application of 60 kg P/ha;
Sufficiency ranges at the uppermost threefoliate leaves R2 stage: 0.26-0.50 % P and 1.71-2.50 % K

Table 20. Response of soybean to P and K fertilization (Casanova, 2000)

Casanova (2000) reported that the primary nutritional limitation for successful soybean production under savanna soils in Venezuela   (Guarico state) are soil acidity and deficiencies of P, K, N and Ca. Other nutrients as Mg, S and Zn become limiting when the soil is cultivated for several years. Application of triple superphosphate up to 70 kg P/ha and potassium chloride up to 135 kg K/ha on fixed rate of 108 kg K/ha for P plots and 60 kg P/ha for K plots resulted by considerable yield increases. The treatment combination of 60 kg P/ha and 108 kg K/ha produced the best grain yield of 3.16 t/ha (Table 20).

Mottaghian et al. (2008) applied in a silty loam soil in Mazandaran province, Iran, for soybean eight fertilization treatments as follows: 20 and 40 t/ha of organic fertilizers (municipal solid waste compost, vermicompost and sewage sludge) enriched with 50% of anorganic fertilizers need by the soil), only inorganic fertilizers (potassium sulphate and triplephosphate 75 kg/ha) and control (unfertilized). Mixture of 40 t/ha sewage sludge and inorganic fertilizers produced the highest yield and micronutrient (Mn, Cu, Zn and Fe) grain concentrations.

9.2 Liming

Acid soils occupy 3.95 billion ha (about 30 % of the world's ice-free land area (von Uexkull and Mutert, 1995). The poor production of crops grown in acid soils is due to combinations of toxicity (Al, Mn, Fe, H) and deficiencies (N, P, Ca, Mg, K, Fe, Zn). Soil acidity is certainly one of the most damaging soil conditions affecting the growth of most crops. Many factors are involved, but Al toxicity is of outermost significance because of damaging root growth and therefore reduces water and nutrient uptake.

Poor growth of soybean in acid soils as been attributed to a number of factors that include: low pH, high level of Al, Mn, and H, low levels of Ca, Mg, P, K, micronutrients like B, Zn etc. (Fageria, 1994), low population of beneficial micro-organisms like rhizobia, vesicular arbuscular (VAM) fungi and inhibition of root growth (Maddox and Soileux, 1991).

Management practices, such as acidificatying effects of acid-forming N fertilizers, removal of cations by harvested crops, increased leaching and leguminous crops (N_2-fixation), have resulted in the lowering of natural soil pH (Baligar and Fageria, 1997).

Plant growth in mineral acid soils can be restricted by complex of factors. Malnutrition of plants on acid soils is mostly the results of limited soil nutrient availability, often strengthened by impaired uptake capability of the root. Based for over 5000 observations for soybeans worldwide, the optimum pH value for soybeans indicated by this approach lies between ph 5.7 and 6.0 (Sumner, 1997).

An optimum liming regime should achieve the reduction plant available Al and Mn concentrations to levels which allow optimal production of a particular crops, supplying adequate levels of plant available Ca and Mg for optimum root growth and crop performance, creating conditions for optimal performance of beneficial soil fauna and flora particularly int he rhizosfere, and in case of legumes, creating environment which promotes infection and nodulation of root with effective N-fixing rhizobia. (Keltjens, 1997).

For successful soybean production, large quantities of lime and phosphorus fertilizers may be required (Fageria et al., 1995). Liming improves microbiological activities of acid soils, which in turn increases N fixation by legumes, and also promotes mineralization of organic materials. However, over liming may reduce crop yield by inducing P and micronutrient deficiencies (Fageria, 1984).

Unfortunately, over 50% of the world's potential arable land surface is composed of acid soils mostly distributed in developing countries (von Uexküll and Mutert, 1995; Kochian et al., 2005). This restricts the production of soybeans and other legumes due to their sensitivity to acid soil infertility. The growth of leguminous crops and development of symbiosis on acid soils are generally affected by deficiencies of Ca, K, P, Mg, S, Zn and Mo and/or toxicities of Al, Mn and Fe (Foy,1984; Clark et al., 1988).

Liming has been used to ameliorate the problem of aluminium toxicity and low pH in soils. Liming the top soil, however, remains a temporary solution due to subsoil acidity. Restriction in root growth due to subsoil acidity reduces plant nutrient acquisition and access to subsoil water which culminates in the reduction of crop yield (Ferrufino et al., 2000). Moreover, the cost of liming particularly in developing countries is prohibitive and does not justify such huge investment given the return on investment from grain yield of soybeans. The input cost of the recommended quantity of 0.5 to 1.00 tons/ha of liming material (Yusuf and Idowu, 2001), is about the expected total revenue from the current average yield of 0.7 tons/ha in the South-East and South-South regions of Nigeria. The identification of acid stress tolerant cultivars of soybeans, therefore, remains a viable alternative.

Opkara et al., (2007) conducted field experiments in Southeastern Nigeria, in the 2003 and 2004 growing seasons to assess the effect of liming on the performance of four high yielding soybean varieties (early maturing TGX 1485-1D, TGX 1799-8F, TGX 1805-8F and medium maturing TGX 1440-1E). Five lime rates of 0, 0.5, 1.0, 1.5 and 2.0 t/ha were applied to the main plots while the four soybean varieties were planted in the sub-plots. Liming significantly increased soil pH, number of nodules and number of pods per plant and grain yield, especially in 2004 but did not significantly influence plant height, shoot dry matter, days to 50% flowering and 100-seed weight. The 1.0 t/ha lime rate proved to be optimum and is thus recommended for high grain yield in soybean. Mean grain yield at 1.0 t/ha lime rates was higher than the yield in the control (no lime) by 66%. The medium maturing TGX 1440-1E gave, on the average, significantly higher number of leaves and number of pods per plant and grain yield than other varieties.

Kovacevic et al., (1987) tested response of maize, soybean and wheat on liming by hydrated lime to level 20 t/ha. The field experiment was conducted in triplicate for maize- soybean-wheat rotation. Depending on the year grain yield of soybean ranged between 2.59 and 4.03 t/ha. Liming with 10 t of lime increased soybean yield by 17 %. Increased lime rates did not affect grain yield. Low grain yield were obtained under dry and warm weather conditions in 1983 and cold and wet weather conditions in 1984. Liming with 20 tons of lime per hectare increased soil pH from 4.0 to 6.4 at end of the first year of testing (Table 21).

	Hydrated lime (t/ha: autumn 1980)						Mean	LSD	
Year	0	1	5	10	15	20		5%	1%
	Grain yield of soybean in maize-soybean-wheat rotation (t/ha)								
1981	2.67	2.72	2.80	2.81	2.95	2.85	2.80	0.11	0.15
1982	3.48	3.75	4.06	4.28	4.29	4.32	4.03	0.11	0.15
1983	2.24	2.39	2.60	2.72	2.77	2.83	2.59	0.08	0.11
1984	2.27	2.32	2.49	3.08	3.03	3.02	2.70	0.15	0.20
1985	3.37	3.43	3.58	3.61	3.59	3.59	3.53	0.09	0.12
Mean	2.81	2.92	3.11	3.30	3.33	3.32			
	Soil pH (1n KCl) status at end of the growing season								
1981	4.03	4.21	5.03	5.44	5.94	6.42			
1984	4.40	4.36	4.93	5.46	6.02	6.53			

Table 21. Response of soybean to liming on Fericanci acid soil (Kovacevic et al., 1987)

Loncaric et al. (2007) applied liming with carbocalk up to 20 t/ha (spring 2003) and three degrees of fertilization (every year for 4-year period) on Donji Miholjac dystric luvisol. Soybean was grown on the experimental field in the fourth growing season (2006). Depending on the treatment, soybean yields were in range from 2.7 t/ha (unlimed and unfertilized plots) to 4.4 t/ha (treatment kg/ha: 140 N + 300 P_2O_5 + 300 K_2O) and phosphorus removals (P_2O_5/ha) by soybean were 56 and 78, respectively.

9.3 Soil tillage

Different tillage techniques affect the root absorption of nutrients. Lavado et al. (2001) tested effects of conventional and zero tillage (CT and ZT) on nutritional status of soybean, wheat and maize with emphasis on heavy metals. The field experiments were conducted in area far from contaminated sources in Buenos Aires Province, Argentina. The effects of tillage were limited for nutrient concentrations, but significant for heavy metals. Soybean appeared to be more sensitive than cereals to the apparent effect of soil tillage. Grain composition of soybean was independent on soil tillage. Under CT conditions leaves and stem N as well as root Cu were significantly higher, while root -Zn, -Pb and -Ni were significantly lower in comparison with ZT (Table 22).

Soil tillage treatments (ZT = zero tillage; CT = conventional tillage) and soybean nutritional status (G = grain; L+S = leaves and stems; R = root) of soybean under field conditions (mg/kg)*											
	ZT	CT	ZT	CT	ZT	CT	ZT	CT	ZT	CT	ZT	CT
	Nitrogen		Phosphorus		Potassium		Sulfur		Copper		Zinc	
G	33800a	46900a	4500a	1900a	19800a	18000a	2200a	2600a	17.10a	20.83b	44.85a	43.50a
L+S	9500a	14400b	2100a	1500a	15000a	15000a	900a	800a	10.93a	13.45a	21.48a	18.70a
R	7000a	8600a	2900a	900b	1300a	900a	800a	800a	18.70a	27.98b	64.73a	41.78a
	Boron		Molybdenum		Lead		Nickel		Cadmium		Chromium	
G	5.77a	6.15a	2.95a	1.71a	0.85a	0.80a	4.30a	4.26a	<0.05a	<0.05a	0.93a	1.20a
L+S	4.03a	4.60a	1.49a	1.70a	0.69a	0.63a	1.55a	2.08a	<0.05a	<0.05a	1.74a	2.36a
R	5.10a	6.00a	1.15a	1.36a	3.51a	2.41b	9.46a	6.77b	<0.05a	<0.05a	10.80a	11.93a
	* Means with different letter in each row are significantly different between treatments at LSD 5 %											

Table 22. Impacts of soil tillage on nutrient and heavy metal status of soybean (Lavado et al., 2001)

Jug et al (2006) reported about soil tillage impacts on nutritional status of soybean on chernozem soil for four growing seasons (stationary field experiment from 2002 to 2005). Three treatment of soil tillage were applied as follows: a) conventional tillage, b) reduced tillage (DH = diskharrowing instead of ploughing) and c) no-till (NT). In general, the characteristics of growing season (the factor „year") were more influencing factor of soybean nutritional status (aerial part in stage of full-developed pods) in comparison with the soil tillage. In this study, low influences of applied soil tillage treatments on nutritional status of soybean were found because significant differences on soybean composition were found only for four (Cu, Cr, Sr and Ba) from total 20 analysed elements. For example, conventional tillage resulted by the higher plant Cu (by 15% and 18% in comparison with DH and NT, respectively), and the lower plant Sr (by 12% and 16%, respectively) and Ba (by 26% and 23%, respectively), while under DH conditions by 22% lower plant Cr was found. Main nutrient status were independent on soil tillage (Table 23). For this reason, usual fertilization practice is recommended for possible application of soil tillage reduction under conditions of calcareous chernozem.

Stipesevic et al. (2009) reported response of winter wheat and soybean to different soil tillage systems on chernosem soil for four years. Three applied soil tillage treatments were applied as follows: a) CT – conventional soil tillage, based on mouldboard ploughing, b) DH – soil tillage based on diskharrowing instead of ploughing; and c) NT – no-tillage. Both crops showed decreasing concentration of Zn within the plant tissue as a result of the soil tillage reduction in the order CT>DH>NT, presumably due to the limited roots growth in lesser disturbed soil at DH and NT treatments. Winter wheat recorded generally lower than optimal Zn concentrations and higher P:Zn ratios at reduced soil tillage treatments, as a result of lower Zn uptake. The recommendation for the winter wheat production by reduced soil tillage is additional Zn fertilization, whose exact amounts and way of application shall follow further research.

The year (the factor A) and soil tillage (ST = the factor B: conventional = CT; diskharrowing = DH; no-till = NT) and composition of soybean (cultivar Tisa)**										
Year (A)	ST (B)	The aerial part of soybean at full-developed pods stage*								
		Percent on dry matter basis					mg/kg on dry matter basis				
		P	K	Ca	Mg	S	Zn	Mn	Fe	Cu	B
2002		0.316	1.85	1.67	0.636	0.171	21.9	38.6	158	9.4	37.6
2003		0.297	1.56	1.91	0.639	0.212	22.4	111.1	385	8.7	46.7
2004		0.407	2.49	1.75	0.470	0.265	34.3	69.2	229	10.3	59.2
2005		0.397	2.66	1.74	0.515	0.223	24.0	84.3	282	9.5	44.2
LSD A 5%		0.047	0.47	n.s.	0.101	0.034	3.7	26.5	127	n.s.	6.0
	CT	0.370	2.18	1.68	0.538	0.226	25.3	71.7	272	10.5	46.8
	DH	0.366	2.31	1.75	0.565	0.215	26.7	75.0	222	9.1	47.3
	NT	0.327	1.93	1.87	0.592	0.212	25.0	80.7	296	8.9	46.7
LSD B 5%		n.s.	n.s.	n.s.	n.s.	n.s.	n.s.	n.s.	n.s.	1.3	n.s.
Mean		0.354	2.14	1.77	0.565	0.218	25.7	75.8	264	9.5	46.9
		mg/kg on dry matter basis					mg/kg on dry matter basis				
		Mo	Co	Ni	Cr	Sr	Ba	Al	Pb	Cd	Na
2002		0.148	0.055	2.81	0.851	25.0	6.59	112	0.376	0.030	44.6
2003		0.198	0.160	1.83	0.609	21.1	6.64	312	0.336	0.042	29.3
2004		0.126	0.163	1.84	0.406	16.4	5.30	171	0.307	0.028	34.8
2005		0.313	0.174	2.44	0.450	21.3	9.48	215	0.503	0.086	59.2
LSD A 5%		0.090	0.040	n.s.	0.193	4.1	2.52	123	0.097	0.034	9.6
	CT	0.267	0.149	2.56	0.626	19.1	5.75	214	0.374	0.051	39.5
	DH	0.175	0.120	2.14	0.485	21.7	7.80	163	0.394	0.043	41.6
	NT	0.146	0.145	1.99	0.626	22.1	7.46	230	0.374	0.045	44.9
LSD B 5%		n.s	n.s.	n.s.	0.126	2.5	1.70	n.s.	n.s.	n.s.	n.s.
Mean		0.196	0.138	2.23	0.579	21.3	9.48	215	0.381	0.047	42.0
* under detectable levels (mg/kg): Se (<0.60), Hg (<0.12), As (<0.40)										

Table 23. Influences of the growing season and soil tillage on nutritional status of soybean (Jug et al., 2006)

10. Mineral nutrition in function of diseases and pest control

To control diseases and pest the farmers have several options as follows: genetics (cultivation of less susceptible or even resistant to diseases and pest), biological control (utilization of predators), chemical control (using correspondingly pesticides), plant and soil management practices (creating optimal growth conditions of the cultivated crops and /or to eradicate those conditions, which are favorable for multiplication of diseases and pest) and plant nutrition.

Nutrition of plant has a substantial impact on the predisposition of plants to be attacked or affected by diseases and pests. The ratio between nitrogen and potassium plays obviously a particular role in the host/pathogen relationship. However, unbalanced fertilization is wead spread Developing countries apply nitrogenous and potassic fertilizers at a ratio of 1: 0.2, the situation in developed countries is slightly better with a NK ratio of 1:0.4 (Krauss, 2001). Generally, potassium tends to improve plant health (Perrenoud, 1990).

Useof potassium decreased the incidence of fungal diseases in 70% of the cases. Simultaneously, K increased yield of plants infested with fungal diseases by 42% (Perrenoud, 1990). Mondal et al. (2001) found a negative correlation between K contents in soybean with incidence and positive correlation with their respective yield.

Insects actively select plants best suited as a food source by, among other factors, appearance, stage of development and composition oft he plant. A precondition for successful infestation is the coincidence of certain developmental stages of both host and pathogen. The use of fertilizers can affect this coincidence by either accelerating or slowing down the development of the host plant relative to that of pathogen. A good example is the control of stem cancer (*Diaporthe phaseolorum*) in soybean by potash use, because the fungus can attack soybean only at a particular phenological stage. Earliness due to balanced fertilization provides the possibility to escape (Ito et al., 2001).

Rodrigues et al., (2009) found that spraying of soybean by potassium silicate (Psi) solutinon reduced the intensity of soybean rust. Soybean rust severity at the highest applied KSi rate in level 60 g/L (pH 5.5) was 70% less than the control (plant spraying with water). This finding may be valuable in areas where soybean is grown as a monoculture, and where high yielding but susceptible cultivars cannot be grown because of occurence of frequent severe epidemics. However, Duarte (2009) reported that there was no effects of KSi on rust control in susceptible soybean cultivar *Monarca*.

11. References

Anetor, M. O. & Akinrinde, E. A. (2006). Response of Soybean [*Glycine max* (L.) Merrill] to Lime and Phosphorus Fertilizer Treatments on an Acidic Alfisol of Nigeria. *Pakistan Journal of Nutrition*, Vol. 5, No. 3, pp. 286-293, ISSN 1680-5194

Arao, T., Ae, N., Sugiyama, M. & Takahashi, M. (2003). Genotypic differences in cadmium uptake and distribution in soybeans. *Plant and Soil*, Vol. 251, pp. 247-253.

Baligar,V. C. & Fageria, N. K. (1997). Nutrient use efficiency in acid soil, nutrient management and plant use efficiency. *Plant-Soil Interactions at Low pH*, A.C. Moniz et al. (Ed.), pp. 75-95, Brazilian Soil Science Society, Campinas/Vicosa.

Barbagelata, P., Melchiori, R. & Paparotti, O. (2002). Phosphorus feertilization of soybeans in clay soils of Entre Rios province. *Better Crops International*, Vol. 16, No 1, pp. 3-5.

Bergman, W. (1992). *Nutritional disorders of plants - development, visual and analytical diagnosis*, Gustav Fischer Vela Jena, Stuttgart, New York.

Bergman, W. & Neubert, P. (1976). *Pflanzendiagnose und pflanzenanalyse*, WEB Gustav Fisher, Jena.

Bethlenfalvay, J. K., Franson, L. R. & Brown, S. M. (1990). Nutrition of mychorrhizal soybean evaluated by the diagnosis and recommendation integrated system (DRYS). *Agronomy Journal*, Vol. 82, pp. 302-403.

Bezdicek, D.F., Evans, D.W., Adebe, B. & Witters, R.E. (1978). Evaluation of peat and granular inoculum for soybean yield and N fixation under irrigation. *Agronomy Journal*, Vol. 70, pp. 865-868.

Borkert, C.M., & Cox, F.R. (1999). Effects of acidity at high soil zinc, copper, and manganese on peanut, rice, and soybean. *Commun Soil Sci Plant Anal*, Vol. 30, pp. 1371-84.

Brown, J.C. & Jones, W.E. (1976). A technique to determine iron efficiency in plants. *Soil Sci. Soc. Am. J.*, Vol. 40, pp.398-405.

Casanova, E. F. (2000). Phosphorus and potassium fertilization and mineral nutrition of soybean in Guarico state. *Better Crops International*, Vol. 14, No 2, pp. 6-9.

Clark, R.B., Flores, C.I. & Gourley, L.M. (1988). Mineral element concentrations in acid soil tolerant and susceptible Sorghum genotypes. *Communications in Soil Science and Plant Analysis*, Vol. 19, pp. 1003 – 1017.

Campo, R. J., Araujo, R. S. & Hungria, M. (2009). Molybdenum-enriched soybean seeds enhance N accumulation, seed yield, and seed protein content in Brazil. *Field Crops Research*, Vol. 110, pp. 219-224.

Chaney, R.L. & Coulombe, B.A. (1982). Effect of phosphate on regulation of Fe-stress response in soybean and peanut. *Journal of Plant Nutrition*, Vol. 5, pp. 469-487.

Chen, H.M., Zheng, C.R., Tu, C. & She, Z.G. (2000). Chemical methods and phytoremediation of soil contaminated with heavy metals. *Chemosphere*, Vol. 41, pp. 229–34.

Coulombe, B.A., Chaney, R.L. & Wiebold, W.J. (1984). Bicarbonate directly induces iron chlorosis in susceptible soybean cultivars. *Soil Sci. Soc. Am. J.*, Vol. 48, pp. 1297-1301.

Council for Agricultural Science and Technology (2009). *Sustainable of U.S. Soybean Production*. Conventional, Transgenic and Organic Production Systems. Special publication No 30, Ames, Iowa, USA, pp. 24-30.

Cox, F.R. (1965). Factors affecting the Mn content and yield of soybeans. *NC Soil Sci Soc Proc*, Vol. 8, pp. 57-67.

Cox, F.R. (1968). Development of a yield response prediction and manganese soil test interpretation for soybeans. *Agronomy Journal*, Vol. 60, pp. 521-524.

Dahnke, W. C., Fanning, C. & Cattanach, A. (1992). Fertilizing soybean, Available from: www.ag.ndsu.edu/pubs/plantsci/soilfert/sf719w.htm (10. 03. 2011).

Deibert, E.J., Bijeriego, M. & Olson, R.A. (1979). Utilization of ^{15}N fertilizer by nodulating and non-nodulating soybean isolines. *Agronomy Journal*, Vol. 71, pp. 717-723.

De Mooy, W. C., Pesek, J. & Spaldon, E. (1973). Mineral Nutrition. In: *Soybeans. improvement, production, and uses* , Caldwell, B.E. (Ed.), p.267-352, Madison.

Duarte, H. S. S. (2009). Potassium silicate, acilbenzolar-S-methyl and fungicides on the control of soybean rust. *Cienc. Rural* (online), Vol. 39, No 8, pp. 2271-2277. ISSN 0103-8478. Available from http://www.scielo.br (8. 03. 2011.)

Epstein, E. (1976). Genetic potentials for solving problems of soil mineral stress, adaptation of crops to salinity. In: *Plant Adaptatation to Mineral Stress in problem Soils*. M. J. Wright, pp.73-82, Cornell Univ. Agric. Exp. Station, Ithaca NY.

FAOSTAT (2009). Statistics database. Available from htpp.//faostat.fao.org/default.aspx (verified 23 June 2009).

Fageria, N.K., Zimmermann, F.J.P. & Baligar, V.C. (1995). Lime and P interaction on growth and nutrient uptake by upland rice, wheat, common bean and corn in an oxisol. *Journal of Plant Nutrition*, Vol. 18, pp. 2519-2532.

Fageria, N.K. (1984). Response of rice cultivars to liming in certado soil. *Pesq. Agro spec., Bras. Brasilia,* Vol. 19, pp. 883-889.

Fageria, N.K. (1994). Soil acidity affects availability of Nitrogen, Phosphorus and Potassium. *Better crops International,* Vol. 10, pp. 8-9.

Ferrufino, A., Smyth, T.J., Israel, D.W. & Carter Jr. T.E. (2000). Root elongation of soybean genotypes in response to acidity constraints in a subsurface solution compartment. *Crop Science,* Vol. 40, pp. 413 – 421.

Fleming, A.L., Chaney, R. L. & Coulombe, B. A.(1984). Bicarbonate inhibits Fe-stress response and Fe uptake -translocation of chlorosis-susceptible soybean cultivars. *Journal of Plant Nutr*ition, Vol. 7, pp. 699-714.

Fontes, R.L.F. (1992). *Zinc toxicity in soybean as affected by plant iron and sulfur* [dissertation]. Raleigh (NC). North Carolina State University.

Foy, C. D. (1984). Physiological effects of hydrogen, aluminum and manganese toxicities in acid soils. In. *Soil Acidity and Liming* (Adams F. ed.) pp. 57-97. Am. Soc. Agron. Madison, Wisconsin.

Fontes, R.L.F. & Cox, F.R. (1995). Effects of sulfur supply on soybean plants exposed to zinc toxicity. *Journal of Plant Nutrition,* Vol. 18, pp. 1893-1906.

Fontes, R.L.F. & Cox, F.R. (1998). Zinc toxicity in soybean grown at high iron concentration in nutrient solution. *Journal of Plant Nutrition,* Vol. 21, pp. 1723-30.

Franzen, D. W. & Richardson, J. L. (2000). Soil factors affecting iron chlorosis int he Red River Valley of North Dakota and Minnesota. *Journal of Plant Nutrition,* Vol. 23, No. 1, pp. 67-78.

Furlani, A. M. C., Furlani, P. R., Tanak, R. T., Mascarenhas, H. A. A. & Delgado, M. D. P. (2002). Variability of soybean germplasm in relation to phosphorus uptake and use efficiency for dry matter yield. In: *Plant nutrition – food security and sustainability of agro-ecosystems,* Horst W. J. et al., Developments in Plant and Soil Sciences, Vol. 92, pp. 74-75. Kluwer Academic Publishers Dordrecht/Boston/London.

Galloway, J.N., Schlesinger, W.H., Levy, H., Micaels, A. & Schnoor, J.L. (1995). Nitrogen fixation: Atmospheric enhancement-environmental response. *Global Biogeochem Cycles,* Vol. 9, pp. 235-252.

Ganesh, K.S., Baskaran, L., Chidambaram, A.A. & Sundaramoorthy, P. (2009). Influence of chromium stress on proline accumulation in soybean (*Glycine max* L. Merr.) genotypes. *Global J. Environ. Res.,* Vol. 3, No 2, pp.106-108.

Gill, M. S., Singh, V. K. & Shukla, A. A. (2008). Potassium Management for Enhancing Productivity in Soybean-based Cropping System. *Indian J. of Fertilizers,* Vol. 4, pp. 13-22.

Gray, C., Pennington, H. D. & Matocha, J. (1982). Identifying and correcting iron deficiency in field crops. *Texas Agric. Ext. Ser.* L-723.

Goos, R. J. & Johnson, B. (2003). Screening soybean varieties for resistance to iron chlorosis. Available from: http://ww.docstock.com/docs/32446064/107. (10. 03. 2011).

Hansen, N. C., Schmitt, M. A., Anderson, J. E. & Strock, J. S. (2003). Iron deficiency of soybean int he Upper Meadwest and associated soil properties. *Agronomy Journal,* Vol. 95, pp. 1595-1601.

Hanumanthappa, M., Hosmani, S. A. & Sreeramulu, K. J. R. (1998). Effect of different levels of phosphorus on yield and yield components of soybean varieties. *Agricultural Science Digest,* Vol. 18, pp. 140-142.

Hanumanthappa, M. , Sreeramulu, K. J. R. & Naik, R. G.(1999). Influence of phosphorus levels on dry matter production and yield in soybean varieties. *Journal of Maharashtra Agricultural Universities*, Vol. 23, pp.195-196.

Haq, M. U. & Mallarino, A. P. (2005). Response of Soybean Grain Oil and Protein Concentrations to Foliar and Soil Fertilization. *Agronomy Journal*, Vol. 97, pp. 910-918.

Harper, J. E. & Gipson, A. H. (1984). Differential nodulation tolerance to nitrate among legume species. *Crop Science*, Vol. 24, pp. 797-801.

Holshouser, D. (1997). Micronutrient deficiencies. Available from http://webipm.ento.vt.edu/ipm-www/nipmn/VA-IPM/updates/micronu.html. (10. 03. 2011)

Hrustic, M., Vidic, M. & Jockovic, Dj. (1998). *Soja / Soybean*. Institut za ratarstvo i povrtarstvo Novi Sad, Soja protein d.d. Becej, Novi Sad - Becej, Serbia.

Hungria, M. & Campo, R.J. (2004). Economical and environmental benefits of inoculation and biological nitrogen fixation with soybean:situation in South America. *Proceedings of World Soybean Research Conference VII*, pp.488-498, ISBN 85-7033-004-9, Brasil, March 2004, Foz do Iguassu.

Ito, M. F., Mascayrenhas, H. A. A., Tanaka, R. T., Martins, A. L. M., Otsuk, I. P., Carmello, Q. A. C. & Muraoka T. (2001). Control of stem cancer in soybeans by liming and potassium fertilizer. *Rev. de Agricultura, Piracicaba*, Vol. 76, fasc. 2, pp. 307-316.

Johnson, J. (1984). Fertilizer reduces yield loss from weather stress. *Potash Review*, Subject 6, Suit 49, No 6, pp.1-3.

Johnson, J. (1987). Soil fertility and crop nutrition. In: *The soybean in Ohio*. Ohio Cooperative Extension Service, The Ohio State University, pp. 34-41.

Jug, I., Jug, D., Kovacevic, V., Stipesevic, B. & Zugec, I. (2006). Soil tillage impacts on nutritional status of soybean. *Cereal Research Communications*, Vol. 34, No 1, pp. 537-540.

Kastori, R. (1978). Change of different forms of Ca in soybean varieties during growth season. *Zbornik za prirodne nauke*, Matica srpska Novi Sad, 54, 85-93 (in Serbian).

Kastori, R., Belic, B., Molnar, I., Petrovic, N. & Dzilitov, S. (1979). Dynamics of content, accumulation and distribution of N, P, K, Ca and Mg during growth of some soybean varieties. *Savremena poljoprivreda*, Novi Sad, Vol. 27, pp. 433-446. (in Serbian)

Keltjens, W. G. (1997). Plant adaptation and tolerance to acid soil; it is possible Al avoidance - a review. In: *Plant-Soil Interactions at Low pH*, A.C. Moniz et al. (Ed.), p. 109-11. Brazilian Soil Science Society Campinas/Vicosa.

Keogh, J., Sable, W. E. & Caviness, C. E. (1977). Leaf nutrient concentration in selected soybean cultivars as affected by fertilization, stage of growth and year. *Report Series, Agr. Exp. Station, University of Arkansas*, N⁰ 234.

Keyser, H.H. & Li, F. (1992). Potential for increasing biological nitrogen fixation in soybean. *Plant and Soil*, Vol. 141, pp. 119-135.

King, L.D. (1986). Agricultural use of municipal and industrial sludges in the southern United States. Raleigh (NC). *North Car olina State University. Southern cooperative series; Bulletin 314*.

King, L.D. & Hajjar, L.M. (1990). The residual effect of sewage sludge on heavy metal content of tobacco and peanut. *Journal Environ Quality*, Vol. 19, pp. 738-48.

Kovacevic, V. (1993). Phosphorus and potassium deficiency in soybeans on gleysol of Croatia. *Rostlinna Vyroba*, Vol. 39, No 4, pp. 345-352.

Kovacevic, V. & Basic, F. (1997). The soil potassium resources and the efficiency of potassium fertilizers in Croatia (Country Report 10). *International Potash Institute, Coordinator Central/Eastern Europe*, CH-4001 Basel/Switzerland.

Kovacevic, V. & Grgic, D. (1995). Response of soybeans to potassium fertilization on a high potassium fixing soil. *Rostlinna Vyroba*, Vol. 41, No 4, pp. 24-248.

Kovacevic, V., Juric, I., Zugec, I. (1987). Soybean response to lime and phosphorus under the growing conditions of eastern Croatia. *Eurosoya* Vol. 6. pp. 58-61.

Kovacevic, V. & Krizmanic, M. (1987). Nutrient uptake by soybean genotypes on calcareous soil in eastern Croatia. *Eurosoya*, Vol. 6, pp. 71-73.

Kovacevic, V., Seput, M., Andric, L. & Sostaric, J. (2007). Response of maize and soybeans to fertilization with phosphorus and potassium on acid soil. *Cereal Research Communications*, Vol. 35, No. 2, pp. 645-648

Kovacevic, V., Sudaric, A., Sudar, R., Rastija M. & Iljkic D. (2011). Residual impacts of liming and fertilization on soybean yield and quality. *Növénytermelés*, Vol. 60, No.2 (suppl.), pp.259-262.

Kovacevic, V. & Vukadinovic, V. (1992). The potassium requirements of maize and soybean on a high K-fixing soil. *South African Journal of Plant and Soil*, Vol. 9, pp.10-13.(reprint. *Potash Review*, Subject 6, Cereals, 8th suite, No 1, 1992, International Potash Institute Basel, Schwitzerland).

Kovacevic, V., Vukadinovic, V. & Komljenovic, I. (1991). Tipovi kloroze soje uslijed debalansa ishrane na tlima istočne Hrvatske (The types of soybean chlorosis induced by nutritional debalances on soils of eastern Croatia). *Znan. Prak. Poljopr. Tehnol*, Vol. 21 (Suppl.), pp. 15-23.

Krauss, A. (2001). Potassium and biotioc stress. *Proceedings of 1st FAUBA-FERTILIZAR-IPI Workshop, Potassium in Argenina`s agricultural systems*, November 20-21 Buenos Aires, Argentina.

Lavado, R. S., Porcelli, C. A. & Alvarez, R. (2001). Nutrient and heavy metal concentration and distribution in corn, soybean and wheat as affected by different tillage systems int he Argentine Pampas. *Soil & Tillage Research*, Vol. 62, pp. 55-60.

Li, S., Liu, R.,.Wang, M.,.Wang, X., Shen, H. & Wang, H. (2006). Phytoavailability of cadmium to cherry-red radish in soils applied composed chicken or pig manure. *Geoderma*, Vol. 136, pp. 260–271.

Loncaric Z., Popovic B., Karalic K., Rastija D., Engler M. (2007). Phosphorus fertilization and liming impact on soil properties. *Cereal Research Communications* Vol. 35, No. 2, pp. 733-746.

Maddox, J.S. & Soileux J. M. (1991). Effects ofphosphate fertilizer, lime amendment and innoculation with VA-mychorrhizal fungi on soybean in an acid soil. In: *Plant and Soil interraction at low pH* . Wright *et al.* (Ed.) pp: 215-225. Kluwer Academic Publishers.

Mascagni, H.J. Jr & Cox, F.R. (1985). Critical levels of manganese in soybean leaves at various growth stages. *Agronomy Journal*, Vol. 77, pp. 373-5.

Mascagni, H.J. Jr & Cox, F.R. (1985). Effective rates of fertilization for correcting manganese deficiency in soybeans. *Agronomy Journal*, Vol. 77, pp. 363-366.

Mengel, K. & Kirkby, E. A. (2001). *Principles of Plant Nutrition*. Kluwer Academic Publishers , Dordrecht/Boston/London. ISBN 0-7923-7150-X.

Miner, G.S., Gutierrez, R. & King, L.D. (1997). Soil factors affecting plant concentrations of cadmium, copper, and zinc on sludge-amended soils. *Journal of Environment Quality*, Vol. 26, pp. 989-994.

Mondal, S. S., Pramanik, C. K. & Das, J. (2001). Effects of nitrogen and potassium on oil yield, nutrient uptake and soil fertility in soybean – sesame intercropping system. *Indian J. Agric. SCi.*, Vol. 71, pp. 44-46.

Morshed, R. M., Rahman, M. M., Rahman, M. A. & Hamidullah, A. T. M. (2009). Effect of Potassium on Growth and Yield of Soybean. *Bangladesh J. Agric. and Environ.*, Vol. 5, No 2 , pp. 35-42.

Mottaghian, A., Pirdashti, H., Bahmanyar, M. A. & Abbasian, A. (2008). Leaf and seed micronutrient accumulation in soybean cultivars in response to integrated organic and chemical fertilizers application. *Pakistan Journal of Biological Sciences*, Vol. 11, No 9, pp. 1227-1233.

Nelson, K. A., Motavally, P.P. & Nathan, M. (2005). Response of No-Till Soybean [*Glycine max* (L.) Merr.] to Timing of Preplant and Foliar Potassium Applications in a Claypan Soil. *Agronomy Journal*, Vol. 97, pp. 832-838.

Ogburia, M. N., Atabaeva, H. N. & Hassanshin, R. U. (1999). Evaluation of varietal response of soybean (*Glycine max* L. Merril) to nitrogen (N) fertilization in Tashkent, Central Asia. *Acta Agronomica Hungarica*, Vol. 47, pp. 329-333.

Ojo, G. O. S., Bello, L. L. & Adeyemo, M. O. (2010). Genotypic variation for acid stress tolerance in soybean in the humid rain forest acid soil of south Eastern Nigeria. *Journal of Applied Biosciences*, Vol. 36, pp. 2360- 2366. ISSN 1997–5902. Available from www.biosciences.elewa.org. (10. 03. 2011).

Opkara, D. A., Muoneke, C. O. & Ihediwa, N. (2007). Influence of liming on the performance of high-yielding soybean varieties in southeastern Nigeria. *Journal of Agriculture, Food, Environment and Extension*, Vol. 6, No 2, ISSN 1119-7455 URL: http://www.agrosciencejournal.com/. (11. 03. 2011).

Pandey, S.K. & Pandey, S.K. (2008). Germination and Seedling growth of Field Pea *Pisum sativum* Malviya Matar-15 (HUDP-15) and Pusa Prabhat (DDR-23) under varying level of Copper and Chromium. *J. Am. Sci.*, Vol. 4, No. 3, pp. 28-40.

Perrenoud, S. (1990). Potassium and plant health. *IPI Research Topics No. 3, 2nd rev. Edition*, Basel, Switzerland.

Pichtel, J. & Bradway, D. (2008). Conventional crops and organic amendments for Pb, Cd and Zn treatment at a severely contaminated site. *Bioresource Tech*, Vol. 99, pp. 1242–1251.

Putwattanaa, N., Kruatrachueb, M., Pokethitiyooka, P. & Chaiyaratc, R. (2010). Immobilization of cadmium in soil by cow manure and silicate fertilizer, and reduced accumulation of cadmium in sweet basil (Ocimum basilicum). *ScienceAsia*, Vol. 36, pp. 349-354. doi: 10.2306/scienceasia1513-1874.2010.36.349

Queiroz, E. F., Garcia, A. A., Sfredo, G. J. & Gordeiro, D. S. (1980). Resposta de cultivares de soya a niveis de abudacao fosfatada. *Resultados de pesquisa de soya 1979/80*, Londrina, Brazil, pp. 203-205.

Raper, J. R. & Barber, S. A. (1970). Rooting System of Soybeans. II. Physiological effectiveness as nutrient absorption surfaces. *Agronomy Journal*, Vol. 62, pp. 585-588.

Rastija, M., Kovacevic, V., Vrataric, M., Sudaric, A. & Krizmanic, M. (2006). Response of maize and soybeans to ameliorative fertilization in Bjelovar-Bilogora County. *Cereal Research Communications*, Vol. 34 (part 2), pp. 641-644.

Rehm, G. & Schmitt, M. (1997). Potassium for crop production. (www.extension.umn.edu/distribution/cropsystem/DC6794.html). (08. 03. 2011).

Reid, K. & Bohner, H. (2007). Interpretation of Plant Analysis for Soybeans. Available from: http:// www.omafra.gov.on.ca/english/crops/facts/soybean_analysis.htm.

Richter, D., Kovačević, V. & Flossmann, R. (1990). Ergebnisse und Untersuchung von K-fixierenden Boden Jugoslawiens. *Richtig dungen-mehr ernten*, KALI-BERGBAU Handelsgesellsch. mbH Berlin, Vol. 14, No 2, pp. 1-6.

Rodriguez, F. A., Duarte, H. S. S., Domiciano, G. P., Souza, C. A., Korndorfer, G. H. & Zambolim, G. (2009). Foliar application of potassium silicate reduces the intensity of soybean rust. *Australasian Plant Pathology*, Vol. 38, No 4, pp. 366-372.

Romheld, V. & Marschner, H. (1981). Effect of Fe stress on utilization of Fe chelates by efficient and inefficient plant species. *Journal of Plant Nutrition*, Vol. 3, pp. 551-560.

Rose, I. A., Felton, W. L. & Banks, L. W. (1981). Response of four soybean varieties to foliar zinc fertilizer. *Australian Journal of Experimental Agriculture and Animal Husbandry*, Vol. 21 (109), pp. 236-24.

Russelle, M.P. & Birr, A.S. (2004). Biological nitrogen fixation: Large Scale Assessment of Symbiotic Dinitrogen fixation by Crops: Soybean and Alfalfa in the Mississippi River Basin. *Agronomy Journal*, Vol. 96, pp. 1754-1760.

Sabbe, W. E. & Delong, R. E. (1998). Influence of phosphorus plus potash fertilizer and irrigation on grain yields of soybean cultivars. *Arkansas Experiment Station Research Series* 459.

Sarawgi, S. K. & Tripathi, R. S. (1998). Yield and economics f soybean (*Glycine max* L. Merril) varieties in relation to phosphorus level. *Journal of Oilseeds Research*, Vol. 15, pp. 363-365.

Saric, M. R. (1981). Genetic spewcifity in relation to plant mineral nutrition. *Journal of Plant Nutrition*, Vol. 3, No 5, pp. 743-766.

Saric M. & Krstic B. (1982).Genetisch bedingte Unterschiede im Stickstoffgehalt verschiedener Sojasorten. Archiv fuer Acker- und Pflanzenbau und Bodenkunde, Vol. 26, No 12, pp. 755-761.

Saric, M. & Loughman, B. C. (eds.) (1983). *Genetic aspects of plant mineral nutrition*. Martinus Nijhoff / Dr. W. Junk Publishers, The Hague, The Netherlands.

Sarker, S. K., Chowdhury, M. A. H. & Zakir, H. M. (2002). Sulphur and boron fertilization on yield quality and nutrient uptake by Bangladesh Soybea-4. *Journal of Biological Sciences*, Vol. 2, pp. 729-733.

Sauchelli, V. (1969). *Trace elements in agriculture*. Nostrand Reinhold Company, New York.

Saxena, S. C., Manral, H. S. & Chandel, A. S. (2001). Effects of inorganic and organic sources of nutrients on soybean. *Indian Journal of Agronomy*, Vol. 46, No 1, pp. 135-140.

Scoccianti, V., Crinelli, R., Tirillini, B., Mancinelli, V. & Speranza, A. (2006). Uptake and toxicity of Cr(III) in celery seedlings. *Chemosphere*, Vol. 64, pp. 695-1703.

Sharma, D.C., Chatterjee, C. & Sharma, C.P. (1995). Chromium accumulation by barley seedlings (*Hordeum vulgare* L.). *Journal of Experimental Boany*, Vol. 25, pp. 241-251.

Shute, T. & Macfie, S. M. (2006). Cadmium and zinc accumulation in soybean: A treat to food safety? *Sceince of the Total Environment*, Vol. 371, pp. 63-73. Availabvle from: www.sciencedirect.com. (08. 03. 2011).

Shamsi, H. I., Jiang, L., Wei, K., Jilani, G., Hua, S. & Zhang, G. P. (2010). Alleviation of cadmium toxicity in soybean by potassium supplementation. *Journal of Plant Nutrition*, Vol. 33, pp. 1926-1938.

Silman, Z.T. & Motto, H. L. (1990). Differential response of two soybean genotypes to zinc induced iron deficiency chlorosis. *Journal King Saud. Univ.*, Vol. 2, Agric. Scie (1), pp. 71-80.

Singh, G., Singh, H., Kolar, J. S., Singh, G. & Singh, H. (2001). Response of soybean (*Glycine max* (L) Merrill.) to nitrogen, phosphorus, potassium and zinc fertilization. *Journal of Research Punjab Agricultural University*, Vol. 38, No 1-2, pp. 14-16.

Singh, G. B. & Swariup, A. (2000). Lessons from long-term fertility experiemnts. *Fertilizer News*, Vol. 45, No 2, pp. 13-24.

Slaton, N. A., Golden, B. R., DeLong, R. E. & Mozaffari, M. (2009). Correlation and Calibration of Soil Potassium Availability with Soybean Yield and Trifoliolate Potassium. *Soil Sci. Soc. Am. Journal*, Vol. 74, No. 5, pp. 1642-1651.

Spehar, C.R. (1995a). Genetic differences in the accumulation of mineral elements in seeds of tropical soybeans (*Glycine max* (L.) Merril). *Pesquisa Agropecuária Brasileira*, Vol. 30, pp. 89-94.

Spehar, C.R. (1995b). Diallel analysis for mineral element absorption in tropical adapted soybeans (*Glycine max* L. Merril). *Theoretical and Applied Genetics*, Vol. 90, pp. 707-713.

Spehar, C.R. (1999). Diallel analysis for grain yield and mineral absorption rate of soybeans grown in acid Brazilian Savannah soil. *Pesquisa Agropecuária Brasileira*, Vol 34, pp.1003-1009.

Stipesevic, B., Jug, D., Jug, I., Tolimir, M., & Cvijovic, M. (2008). Winter wheat and soybean zinc uptake in different soil tillage systems. *Cereal Research Communications*, Vol. 37, No. 2, pp. 305–310.

Sugiyama, M. & Noriharu, A. E. (2009). Differences among soybean cultivars with regard to the cadmium-accumulation patterns in various organs. *Proceedings of the XVI International Plant Nutrition Colloquium, UC Davis, California.* Available from http://escholarship.org/uc/item/71n1j212. (08. 03. 2011.)

Sudaric, A., Vratarić, M., Duvnjak, T. Majic, I. & Volenik, M. (2008). The effectiveness of biological nitrogen fixation in soybean linked to genotype and environment. *Cereal Research Communications*, Vol. 36, pp. 67-70.

Sumner, M. E. (1997). Procedures used for diagnosis and correction of soil acidity, a critical review. *Plant-Soil Interactions at Low pH*, A.C. Moniz et al. (Ed.), pp. 195-204. Brazilian Soil Science Society Campinas/Vicosa.

Tordoff, G.M., Baker, A.J.M. & Willis A.J. (2000). Current approaches to the revegetation and reclamation of metalliferous mine wastes. *Chemosphere*, Vol. 41, pp. 219–28

Uebel E. (1999). Znacenje magnezija u povecanju prinosa i kvalitete usjeva – rezultati poljskih pokusa u nekim europskim zemljama / Yield and quality increases by magnesium fertilization – results of field experiments in some European countries. Poljoprivreda Vol. 5 (99), No. 1, pp. 47-53.

Vincent, J.M. (1980). Factors controllling the legume- Rhizobium symbiosis. *Nitrogen fixation*, Vol. 2, Eds. W.E. Newton & W.H. Orme-Johnson, pp. 103-129. University Park Press, Baltimore, MD.

Von Uexkull & Mutert, E. (1995). Global extent, development and economic impact of acid soils. *Plant-Soil Interactions at Low pH: Principles and Management.* (Date R. A., et al. Ed.) pp. 5-19. Kluwer Academic Publisher, Dordrecht, The Netherlands.

Vrataric, M., Sudaric, A. (2008). *Soja / Soybean.* Sveuciliste J.J. Strossmayera u Osijeku i Poljoprivredni institut Osijek, Hrvatska.

Vrataric, M., Sudaric, A., Kovacevic, V., Duvnjak, T., Krizmanic, M. & Mijic, A. (2006). Response of soybean to foliar fertilization with magnesium sulfate (Epsom salt). *Cereal Research Communications,* Vol. 34, No 1, pp. 709-712.

Vukadinović, V., Bertić, B. & Kovačević, V. (1988). Kalium- und Phosphorverfuegbarkeit auf den Boeden in Gebiet von Posavina. *Tagungsbericht - Akademie der Landwirtschaftswissenschaften der DDR,* No. 267, pp.73-80.

Welch, L. F., Boone, L. V., Chambliss, C. G., Christiansen, A. T., Mulvaney, DF. L., Oldham, M. G. & Pendleton, J. W. (1973). Soybean yields with direct and residual nitrogen fertilization. *Agronomy Journal,* Vol. 65, pp. 547-550.

Williams, C.B. (1930). Factors influencing the productivity of muck soils. Raleigh (NC) *North Carolina Agricultural Experiment Station. Report 53.* pp 46-47.

Willis, L.G. (1928). Response of oats and soybeans to manganese on some coastal plain soils. *North Carolina Agric. Exp. Station Bulletin 257.*

Willis, L.G. (1937). Evidences of the significance of oxidation-reduction equilibrium in soil fertility problems. *Soil Science Society of America Proceeding,* Vol. 1, pp. 291-297.

Willis, L.G & Piland, J.R. (1936). The function of copper in soils and its relation to the availability of iron and manganese. *Journal of Agricultural Research,* Vol. 52, pp. 467-76.

Win, M., Nakasathien, S. & Sarobol, E. (2010). Effects of phosphorus on seed oil and protein contents and phosphorus use efficiency in some soybean varieties. *Kasetsart Journal (Nat. Sci.),* Vol. 44 , pp. 1-9.

Yin, X. Y. & Vyn, T. J. (2004). Critical Leaf Potassium Concentrations for Yield and Seed Quality of Conservation-Till Soybean. *Soil Science Society of America Journal,* Vol. 68, pp. 1626-1634.

Yusuf, I.A. & Idowu, A. A. (2001). Evaluation of four soybean varieties for performance under different lime regimes on the acid soil of Uyo, Nigeria. *Tropical Oilseeds Journal,* Vol. 6, pp. 65 – 70.

Zheng, H., Chen, L., Han, X., Ma, Y. & Zhao, X. (2010). Effectivenss of phosphorus application in improving regional soybean yields under drought stress. A multivariate regression three analysis. *African Journal of Agricultural Research,* Vol. 5 (23), pp. 3251-3258.

7

Regulation of Leaf Photosynthesis Through Photosynthetic Source-Sink Balance in Soybean Plants

Minobu Kasai

Department of Biology, Faculty of Agriculture and Life Science, Hirosaki University
Japan

1. Introduction

Plant photosynthesis is the basis for matter production needed for all living organisms. In the future, plant photosynthesis would be more important, since environmental problems such as climatic warming due to increasing environmental CO_2 concentration and problems of food and energy shortages due to increasing populations may be severer (von Caemmerer & Evans, 2010; Raines, 2011). Increasing plant leaf photosynthesis and thereby increasing plant matter production would be expected as a realistic way to resolve the problems. There is, however, a well-known hypothesis that in plants leaf photosynthesis can be down regulated through accumulated photosynthetic carbohydrates in leaf under excessive photosynthetic source capacity, which also means sink limitation, although the detailed mechanism is not clear (see Kasai, 2008). Actually, for example, there is evidence for the excessive photosynthetic source capacity causing down regulation of photosynthesis in crop plants under field conditions (Okita et al., 2001; Smidansky et al., 2002, 2007). Therefore, for the better improvement of leaf photosynthesis in plants, it is important to elucidate the regulatory mechanism for leaf photosynthesis under excessive photosynthetic source capacity and thereby clarify way of the improvement of leaf photosynthesis.

To elucidate the regulatory mechanism for leaf photosynthesis under excessive photosynthetic source capacity, experimental construction of the excessive photosynthetic source capacity is important. Excising sink organs such as pods, fruits or flowers from plant materials is a way to construct excessive photosynthetic source capacity, and it has often been conducted to study the regulatory mechanism of photosynthetic source-sink balance in plants (see Kasai, 2008). However, the way excising sink organs results not directly but indirectly in excessive photosynthetic source capacity by diminishing sink capacity, and can give some damages to plant materials. Recent studies using transgenic plants have shown that overexpression of Calvin cycle enzymes (sedoheptulose-1,7-bisphosphatase and fructose-1,6-bisphosphatase) or leaf plasma membrane CO_2 transport protein increases the leaf photosynthetic rate significantly (Raines, 2003, 2006). Therefore, the use of the transgenic plants with improved higher leaf photosynthetic rate may be useful to study the regulatory mechanism for leaf photosynthesis under excessive photosynthetic source capacity, since the higher photosynthetic rate is likely to result in excessive photosynthetic source capacity. However, it seems difficult to analyze the down regulation of photosynthesis that may hide in the improved photosynthetic rate. Actually, down regulation of photosynthesis that is associated with excessive photosynthetic source capacity has not been analyzed in the transgenic plants with improved higher photosynthetic rate. Exposure to high CO_2 or continuous exposure to light of plant materials is thought as the other way to construct excessive photosynthetic source capacity. It is well known that leaf photosynthetic rate, especially, in C_3 plants does not reach the saturation at the present atmospheric CO_2 concentration and thus the rate increases initially under high CO_2 conditions (Ward et al., 1999). Therefore, in C_3 plants, exposure to high CO_2 is expected to result in excessive photosynthetic source capacity. However, the way of exposure to high CO_2 may be not suitable to analyze the regulatory mechanism for leaf photosynthesis under excessive photosynthetic source capacity, because of the same reason described for the transgenic plants with improved photosynthetic rate and well-known action of high CO_2 to decrease stomatal aperture (Bredmose & Nielsen, 2009). In contrast, continuous exposure to light of plant materials, which prolongs photosynthetic period, can result in excessive photosynthetic source capacity without affecting directly the sink organs, leaf photosynthetic rate and stomatal aperture and giving direct damage to the plant materials. Soybean plants, although it is single-rooted soybean leaves, have largely contributed to study the regulatory mechanism for leaf photosynthesis under excessive photosynthetic source capacity through the experimental system using continuous exposure to light. Single-rooted soybean leaves are source-sink model plants with a simple organization of a leaf, a short petiole and roots developed from the petiole in individuals and were developed by Sawada et al. (1986) using the primary leaves of intact soybean plants (*Glycine max* L. Merr. cv. Tsurunoko). Studies using single-rooted soybean leaves have shown that treating the plants with continuous light results in accumulation of photosynthetic carbohydrates (sucrose and starch) in the leaf and decrease in the leaf photosynthetic rate, which correlates with the increase in leaf carbohydrate (sucrose or starch) content (Sawada et al., 1986, 1989, 1990, 1992). Also, it has been shown in the single-rooted soybean leaves that deactivation of Rubisco, a CO_2-fixing enzyme is caused by the treatment of continuous exposure to light (Sawada et al., 1990, 1992). As continuous exposure to light of single-rooted soybean leaves also increased the leaf phosphorylated intermediates' contents (Sawada et al., 1989), and there have been findings that in vitro, inorganic phosphate promotes activation of Rubisco by enhancing the affinity of uncarbamylated inactive Rubisco to CO_2 (Bhagwat, 1981; McCurry et al., 1981; Anwaruzzaman et al., 1995), the studies using single-rooted soybean leaves have suggested that there is a regulatory mechanism of leaf photosynthetic rate through deactivation of Rubisco, which is associated with accumulation of photosynthetic carbohydrates in leaf under excessive photosynthetic source capacity, and that the deactivation of Rubisco may be caused by limitation of inorganic phosphate (Sawada et al., 1990, 1992). Data from a study using single-rooted soybean leaves demonstrate that the plants do not change the leaf area and leaf dry weight other than the weights of major photosynthetic carbohydrates (sucrose and starch) and grow only the roots during experimental period, irrespective of whether light conditions are normal (daily light/dark periods of 10/14 h) or continuous without darkness (Sawada et al., 1986). Although the source-sink model plants with simple source-sink organization have been developed from various plant species, only the single-rooted soybean leaves have been demonstrated to show almost no growth in the source organ (Sawada et al., 2003). No growth of the source organ and the simple organization of source and sink in the single-rooted soybean leaves are attractive characteristics to analyze comprehensively the regulatory mechanism of photosynthetic source-sink balance in plants, including the regulatory mechanism for leaf photosynthesis under excessive photosynthetic source capacity. Actually, as mentioned above, various analyses have been conducted in the single-rooted soybean leaves, especially in studies for elucidating the regulatory mechanism for leaf photosynthesis under excessive photosynthetic source capacity. Therefore, the single-rooted soybean leaves are important plant materials to elucidate further the regulatory mechanism for leaf photosynthesis under excessive photosynthetic source capacity. However, the plants are made artificially, and do not exist in nature, and in addition, as already mentioned, the plant leaf originates from only the primary leaf in intact soybean plants (Sawada et al., 1986). Therefore, there is the possibility that properties of single-rooted soybean leaves may not reflect those of the original, intact soybean plants or the other intact plants. Thus, it is important to examine the regulatory mechanism for leaf photosynthesis under excessive photosynthetic source capacity using the original, intact soybean plants.

The present study used the original intact soybean plants, and it was analyzed how continuous exposure to light affects the leaf photosynthetic rate and related characteristics, such as leaf stomatal conductance and intercellular CO_2 concentration, contents of water, chlorophyll, major photosynthetic carbohydrates (sucrose and starch), total protein and Rubisco protein in leaf, and activity and activation ratio (ratio of initial to total activity) of Rubisco and amount of protein-bound ribulose-1,5-bisphosphate (RuBP) in leaf extract, which were analyzed to evaluate the amount of uncarbamylated inactive Rubisco (Brooks & Portis, 1988). The same series of analyses have not been conducted together in studies that have performed the experiment of continuous exposure to light using plants.

2. Materials and methods

2.1 Plant materials

Soybean (*Glycine max* L. Merr. cv. Tsurunoko) seeds were sown in plastic pots (13.5 cm in height, 12.5 cm in diameter) containing almost equal volumes of vermiculite and sand that had been mixed, and were grown in growth chambers (Koitotoron, HNL type; Koito Industries Ltd., Tokyo, Japan) under daily light/dark periods of 10/14 h, day/night temperatures of 24/17°C and relative humidity of 60 %. After 8 weeks, plants were divided into two groups, and one group was grown for 3 days with continuous light, and another group was grown for 3 days under daily light/dark periods of 10/14 h as controls. Nutrients were supplied once a week with a 1000-fold diluted solution of Hyponex [6-10-5 type (N:P:K = 6:10:5); Hyponex Co., Osaka, Japan], and tap water was supplied in sufficient amounts. Light was supplied with incandescent lamps at an intensity of 480 μmol photons m^{-2} s^{-1} (400-700 nm) at the middle height of plants grown for 8 weeks.

2.2 Leaf photosynthetic rate, stomatal conductance and intercellular CO_2 concentration

Leaf photosynthetic rate, stomatal conductance and intercellular CO_2 concentration were determined in fully expanded fourth trifoliate leaves at a light intensity of 1000 μmol photons m^{-2} s^{-1}, air flow rate of 200 ml min^{-1}, air temperature of 25 °C, relative humidity of 60 % and CO_2 concentration of 350 ppm on day 3 after treating plants with continuous light using a portable photosynthetic analyzer (Cylus-1; Koito Industries Ltd.). After measurements, leaf disks (1.79 cm^2) were cut off from fourth trifoliate leaves, immediately frozen in liquid nitrogen and stored at -80 °C until used for the other analyses described below.

2.3 Other analyses

The activity of Rubisco in leaf extract was determined at 25 °C as described previously (Kasai, 2008). For the initial activity, 20 µl of a leaf extract obtained by homogenizing a leaf disk with ice-cold buffer (100 mM HEPES-KOH, pH 7.8, 2 ml) was added to a cuvette containing 1.98 ml of assay medium [100 mM Bicine-KOH (pH 8.2), 20 mM MgCl$_2$, 20 mM NaHCO$_3$, 5 mM creatine phosphate, 1 mM ATP, 0.2 mM NADH, 20 units creatine kinase, 20 units 3-phosphoglycerate kinase and 20 units glyceraldehyde-3-phosphate dehydrogenase], immediately followed by the addition of RuBP (final concentration 0.6 mM) and mixed well. For total activity, RuBP was added 5 min later after 20 µl of the leaf disk extract was immediately combined with the assay medium. The change in absorbance at 340 nm was monitored using a spectrophotometer (Model U-2000; Hitachi Co., Tokyo, Japan).

The amount of protein-bound RuBP in leaf extract was determined as described previously (Kasai, 2008). A leaf extract (800 µl) obtained by homogenizing a leaf disk with an ice-cold buffer (100 mM HEPES-KOH, pH 7.8, 1 ml) was centrifuged (100 g, 1 min) after loading onto a column containing Sephadex G-50 (bed volume before centrifugation, 4 ml) that had been equilibrated with the same buffer. The eluent (500 µl) from the column lacking free RuBP was centrifuged (10,000 g, 10 min) after mixing with an acidic solution (5.5 M HClO$_4$, 50 µl) to precipitate protein in the eluent. The resulting supernatant was centrifuged (10,000 g, 10 min) after neutralizing to pH 5.6 with K$_2$CO$_3$, and RuBP in the supernatant was determined in the assay medium for determining Rubisco activity using purified spinach Rubisco (0.5 units).

Leaf Rubisco content was determined as described by Makino et al. (1986). Leaf total protein was extracted as described by Makino et al. (1986) and quantified by the method of Bradford (1976). Leaf chlorophyll content was determined according to the method of Mackinney (1941). Leaf sucrose and starch contents were determined as described by Sawada et al. (1999). Leaf water content was analyzed by measuring fresh weight and dry weight of leaf disks. Leaf disks were dried for 2 days at 75 °C.

3. Results

Analyzed leaf photosynthetic rate was significantly lower in intact soybean plants grown for 3 days with continuous light than in control plants grown under daily light/dark periods of 10/14h (Fig. 1).

Leaf stomatal conductance was also significantly lower in continuous light-treated plants than in control plants (Fig. 2). Leaf intercellular CO$_2$ concentration did not differ significantly between control and continuous light-treated plants (Fig. 2).

When activation ratio (percentage of initial activity to total activity) of Rubisco in leaf extract was calculated from analyzed initial and total activities of Rubisco in leaf extract, the ratio was significantly lower in continuous light-treated plants than in control plants (Fig. 3). The ratios in control and continuous light-treated plants were 74.2% and 56.6%, respectively.

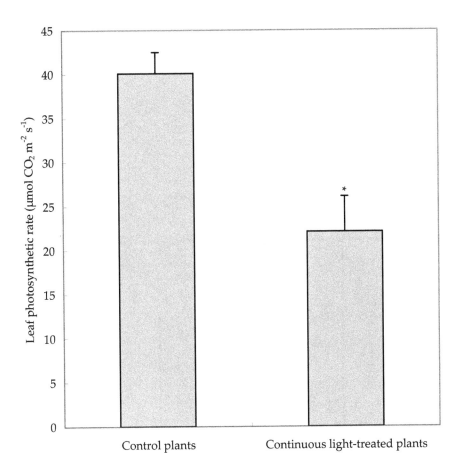

Fig. 1. Leaf photosynthetic rate in soybean plants on day 3 after continuous exposure to light. Control plants were grown under daily light/dark periods of 10/14h for 3 days. Vertical bars indicate S.D. (n=4). *$P<0.01$ (*t*-test) when compared with control plants.

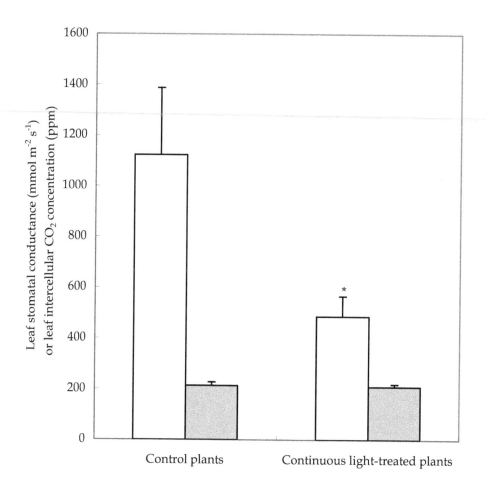

Fig. 2. Leaf stomatal conductance and leaf intercellular CO_2 concentration in soybean plants on day 3 after continuous exposure to light. Control plants were grown as described in Fig. 1. Open bar, leaf stomatal conductance; closed bar, leaf intercellular CO_2 concentration. Vertical bars indicate S.D. (n=4). *$P<0.01$ when compared with control plants. The intercellular CO_2 concentration did not differ significantly ($P>0.05$) between control and continuous light-treated plants.

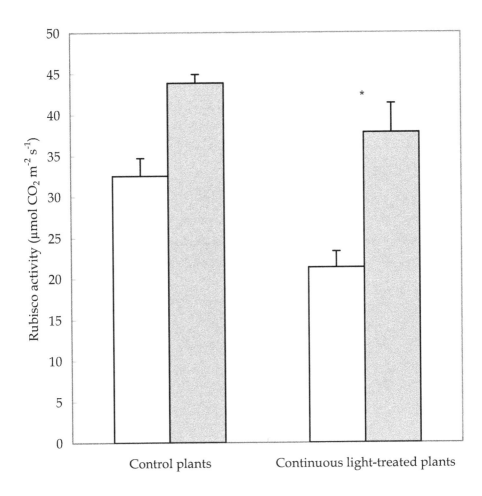

Fig. 3. Initial and total activities of Rubisco in leaf extract from soybean plants on day 3 after continuous exposure to light. Control plants were grown as described in Fig. 1. Open bar, initial activity; closed bar, total activity. Vertical bars indicate S.D. (n=4). In comparison with control plants of the activation ratio of Rubisco calculated as a percentage of the initial activity to total activity, *$P<0.01$.

In a study investigating the light activation of Rubisco using *Arabidopsis thalian*, it was demonstrated that the amount of protein-bound RuBP in leaf extract reflects the amount of uncarbamylated inactive Rubisco (Brooks & Portis, 1988). When the amount of protein-bound RuBP was analyzed, the amount was significantly more in continuous light-treated plants than in control plants (Fig. 4).

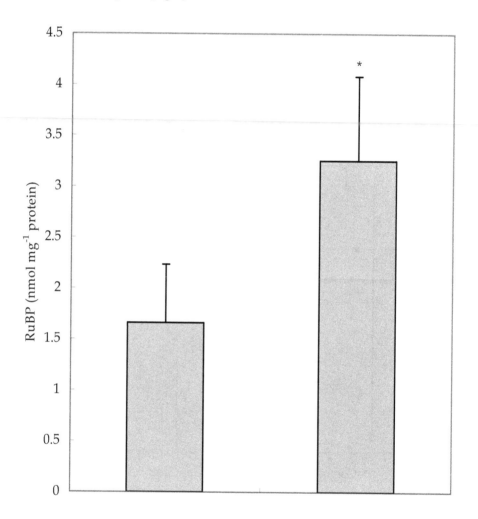

Fig. 4. Amount of protein-bound RuBP in leaf extract from soybean plants on day 3 after continuous exposure to light. Control plants were grown as described in Fig. 1. Vertical bars indicate S.D. (n=4). *$P<0.05$ when compared with control plants.

Contents of sucrose and starch, which are the major photosynthetic carbohydrates, in leaf were both significantly higher in continuous light-treated plants than in control plants (Fig. 5).

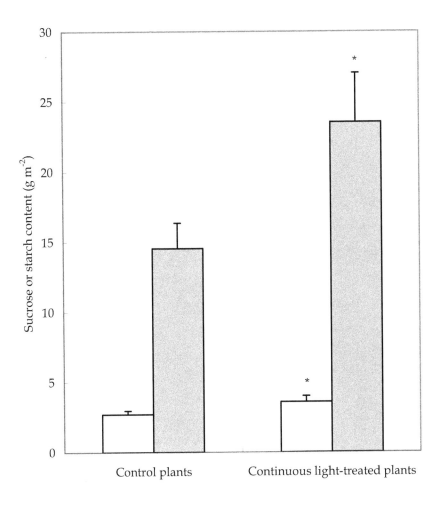

Fig. 5. Leaf sucrose or starch content in soybean plants on day 3after continuous exposure to light. Control plants were grown as described in Fig. 1. Open bar, sucrose content; closed bar, starch content. Vertical bars indicate S.D. (n=4). *P<0.05 when compared with control plants.

Analyzed contents of chlorophyll, water, total protein and Rubisco protein in leaf did not differ significantly between control and continuous light-treated plants (Fig. 6 and 7).

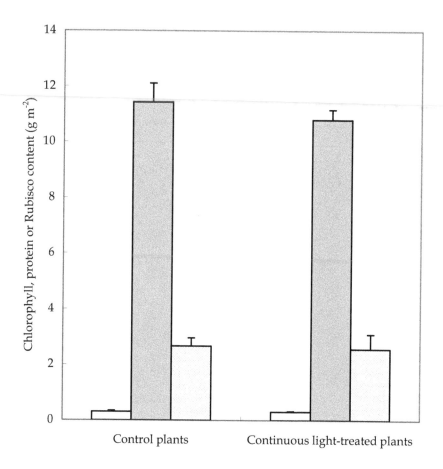

Fig. 6. Leaf chlorophyll, total protein or Rubisco content in soybean plants on day 3 after continuous exposure to light. Control plants were grown as described in Fig. 1. Open bar, chlorophyll content; closed bar, total protein content; dotted bar, Rubisco content. Vertical bars indicate S.D. (n=4). The chlorophyll, total protein and Rubisco contents did not differ significantly (P>0.05) between control and continuous light-treated plants.

Analyzed leaf dry weight other than the weights of sucrose and starch was heavier a little in continuous light-treated plants than in control plants (Fig. 5 and 7). The mean dry weights after subtracting the weights of sucrose and starch in control and continuous light-treated plants were 49.0 g m⁻² and 57.5 g m⁻², respectively.

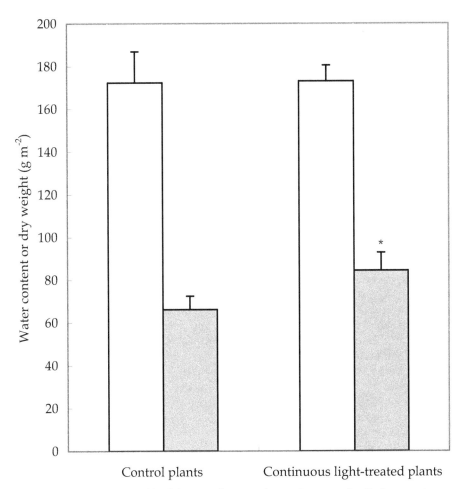

Fig. 7. Leaf water content and leaf dry weight in soybean plants on day 3 after continuous exposure to light. Control plants were grown as described in Fig. 1. Open bar, leaf water content; closed bar, leaf dry weight. Vertical bars indicate S.D. (n=4). *$P<0.05$ when compared with control plants. The leaf water content did not differ significantly ($P>0.05$) between control and continuous light-treated plants.

4. Discussion

The present study was conducted to examine the regulatory mechanism for leaf photosynthesis under excessive photosynthetic source capacity in intact soybean plants. The experimental construction of excessive photosynthetic source capacity was conducted by treating the plants with continuous light for 3 days. The data show that the treatment of continuous exposure to light for intact soybean plants decreased significantly the leaf photosynthetic rate (Fig. 1). Since the light treatment also decreased the leaf stomatal conductance in soybean plants (see Fig. 2), it is thought that the decrease in leaf photosynthetic rate caused by treatment of continuous exposure to light might have resulted from stomatal limitation of CO_2 diffusion. However, the treatment of continuous exposure to light did not affect significantly leaf intercellular CO_2 concentration (see Fig. 2), implicating that the light treatment decreased CO_2 incorporation by leaf photosynthetic cells, as it affected leaf stomatal conductance. In addition, the light treatment decreased activation ratio of Rubisco in leaf extract and did not affect significantly leaf Rubisco content (see Fig. 3 and 6). Furthermore, the light treatment increased the amount of protein-bound RuBP in leaf extract (see Fig. 4). The decrease in activation ratio of Rubisco and increase in the amount of protein-bound RuBP in leaf extract (Brooks & Portis, 1988) strongly suggest an increase in the amount of uncarbamylated inactive Rubisco in leaf. Therefore, it is suggested that the decrease in leaf photosynthetic rate caused by treatment of continuous exposure to light is likely to be due to deactivation of Rubisco in leaf. Treatment of continuous exposure to light for intact soybean plants also increased significantly both the contents of sucrose and starch, which are the major photosynthetic carbohydrates, in leaf (see Fig. 5), indicating that the light treatment could result in an excessive photosynthetic source capacity in the plants. The present study also shows that analyzed leaf chlorophyll, total protein and water contents were not affected significantly by the treatment of continuous exposure to light (see Fig. 6 and 7). Therefore, results obtained in the present study strongly suggest that the decrease in leaf photosynthetic rate in intact soybean plants caused by treatment of continuous exposure to light is unlikely to be due to simple damages such as the breakdown of cellular compartments, but is likely to be due to deactivation of Rubisco, which is associated with accumulation of photosynthetic carbohydrates (sucrose and starch) in leaf under excessive photosynthetic source capacity.

As described in the Introduction, single-rooted soybean leaves have quite been helpful to study the regulatory mechanism for leaf photosynthesis under excessive photosynthetic source capacity, since the plants have simple source-sink organization and have excellent characteristics [growing only the sink organs (roots) without growing source organ (leaf)], which have not been found in other plants (Sawada et al., 1986, 2003). However, as already mentioned, as the plant leaf is constituted from only the primary leaf in intact soybean plants, there is the possibility that properties of single-rooted soybean leaves may not reflect those of the original intact soybean plants or the other intact plants. However, results obtained in the present study of the changes in leaf photosynthetic rate, initial activity and activation ratio of Rubisco in leaf extract, and contents of major photosynthetic carbohydrates (sucrose and starch) and chlorophyll in leaf caused by treatment of continuous exposure to light corresponded with results from studies that have performed similar experiments of continuous exposure to light using single-rooted soybean leaves (Sawada et al., 1986, 1990, 1992). Leaf intercellular CO_2 concentration, amount of protein-bound RuBP in leaf extract and leaf Rubisco content have not been analyzed in the single-rooted soybean leaves. As already mentioned, the present study used the original intact soybean plants from which single-rooted soybean leaves can be made. Therefore, the correspondence of data from original intact soybean plants and those from single-rooted soybean leaves highlights that properties of single-rooted soybean leaves and those of original intact soybean plants are very similar, thus suggesting that properties of single-rooted soybean leaves and those of original intact soybean plants can reflect each other. As described in the Introduction, studies using single-rooted soybean leaves have implicated that there is a regulatory mechanism of leaf photosynthetic rate through deactivation of Rubisco, which is associated with accumulation of photosynthetic carbohydrates in leaf under excessive photosynthetic source capacity (Sawada et al., 1986, 1989, 1990, 1992, 1999, 2003). Data from the present study using the original intact soybean plants have also suggested the same regulatory mechanism of leaf photosynthetic rate. Therefore, the suggested regulatory mechanism of leaf photosynthetic rate may be a common mechanism in plants. With respect to the excellent characteristic of single-rooted soybean leaves that do not change the leaf dry weight other than the weights of major photosynthetic carbohydrates (sucrose and starch) (Sawada et al., 1986), a little change (increase) of leaf (fourth trifoliate leaves) dry weight other than the weights of major photosynthetic carbohydrates (sucrose and starch) was observed by treatment of continuous exposure to light in the original intact soybean plants (see Fig. 5 and 7). Although the present study conducted various analyses to examine the regulatory mechanism for leaf photosynthesis under excessive photosynthetic source capacity, the same series of analyses have not been conducted together in other studies that have performed the treatment of continuous exposure to light using plants.

Treatment of continuous exposure to light for plants results, in most cases, in accumulation of photosynthetic carbohydrate(s) in leaf and decrease in leaf photosynthetic rate. However, in addition to these effects of the light treatment, there are other effects of the light treatment that are different from those indicated by the present study. In tomato, egg plant, peanut and potato, treatment of continuous exposure to light has been shown to result in leaf decolorization (Bradley & Janes, 1985; Globig et al., 1997; Murage et al., 1996, 1997; Rowell et al., 1999; Wheeler & Tibbitts, 1986; Tibbitts et al., 1990). In young leaves of potato and *Arabidopsis*, the continuous light treatment has been shown to accelerate expressions of photosynthetic genes, pigments and proteins, and subsequent declines of the expressions (Cushman et al., 1995; Stessman et al., 2002). In a study using young apple, a decrease in leaf photosynthetic rate caused by treatment of continuous exposure to light was suggested to be due to stomatal limitation of CO_2 diffusion rather than a reduction of Rubisco activity, although, in the study, leaf water content, which is likely to affect stomatal aperture (Brodribb & McAdam, 2011), was not analyzed (Cheng et al., 2004). Therefore, leaf photosynthetic rate may also be regulated through changes in expressions of photosynthetic genes, pigments and proteins and through a regulation of stomata under excessive photosynthetic source capacity in plants.

Other ways, which indirectly construct excessive photosynthetic source capacity as described in the Introduction, have also been shown to result in accumulation of photosynthetic carbohydrate(s) in leaf and decrease in leaf photosynthetic rate. With respect to the cause(s) of why leaf photosynthetic rate declines under the excessive photosynthetic source capacity, for example, data from photosynthetic carbohydrate-feeding or high CO_2 treatment experiments suggest that decreased expressions of photosynthetic genes, including genes for chlororphyll-related protein and Rubisco protein can be causes (Paul & Foyer, 2001; Martin et al., 2002; Paul & Pellny, 2003). However, there is also evidence from high CO_2 treatment experiments using various C_3 plants that decreased Rubisco activity in leaf rather than changes in leaf Rubisco content is likely to be a main cause (Sage et al., 1989). Data from experiments conducting excisions of sink organs (pods or flower buds and flowers) or petiole girdling suggest that a decrease of stomatal conductance or Rubisco activity or Rubisco content in leaf, or both decreases of Rubisco activity and Rubisco content in leaf can be responsible for the decrease in leaf photosynthetic rate under excessive photosynthetic source capacity (Mondal et al., 1978; Setter & Brun, 1980; Setter et al., 1980; Wittenbach, 1982, 1983; Xu et al., 1994; Crafts-Brandner & Egli, 1987; Cheng et al., 2008). As described in the Introduction, excising sink organs or high CO_2 treatment can have side effect(s) other than inducing excessive photosynthetic source capacity. In the present study using intact soybean plants in which excessive photosynthetic source capacity was constructed by treatment of continuous exposure to light, visible damages such as leaf decolorization and wilt were not observed. Treatment of continuous exposure to light did not affect significantly leaf chlorophyll, total protein and water contents analyzed. However, as mentioned above, totally, the effects of indirectly constructed excessive photosynthetic source capacity on leaf carbohydrate status, photosynthetic rate, stomatal conductance, Rubisco activity and photosynthetic gene expressions including Rubisco gene expression are similar to those of excessive photosynthetic source capacity that is constructed by treatment of continuous exposure to light.

Regarding the detailed mechanism(s) of why leaf photosynthetic rate declines under excessive photosynthetic source capacity, recent studies using transgenic plants show that hexokinase could be involved in carbohydrate-mediated repression of photosynthetic gene expression (Jang et al., 1997; Dai et al., 1999; Moore et al., 2003). Other recent study shows that protein kinases (KIN10 and KIN11) may be involved in governing the entirety of carbohydrate metabolism, growth and development in response to carbohydrates in plants (Baena-Gonzalez et al., 2007). Data from a study investigating the effect of chilling stress on leaf photosynthetic rate suggest that H_2O_2, a reactive oxygen species can induce deactivation of Rubisco (Zhou et al., 2006). As described in the Introduction, inorganic phosphate has been found to promote activation of Rubisco by enhancing the affinity of uncarbamylated inactive Rubisco to CO_2 (Bhagwat, 1981; McCurry et al., 1981; Anwaruzzaman et al., 1995). Data from a more recent study suggest that pH within the chloroplasts can be an important factor affecting leaf photosynthetic rate, since the study has demonstrated that pH can affect distribution of Rubisco activase within the chloroplasts by affecting binding of the enzyme to the thylakoid membranes (Chen et al., 2010). Distribution of Rubisco activase within the chloroplasts can affect activation state of Rubisco, since Rubisco activase plays a role in promoting the activation of Rubisco by dissociating RuBP from uncarbamylated inactive Rubisco (Crafts-Brandner & Salvucci, 2000), which tightly binds RuBP (Jordan & Chollet, 1983). Since ATP is needed for the catalytic action of Rubisco activase (Crafts-Brandner & Salvucci, 2000) and it is well known that ATP is needed for regeneration of RuBP, a substrate for Rubisco in Calvin cycle (see Kasai, 2008), it is evident that ATP is also an important factor affecting leaf photosynthetic rate. However, the precise mechanism of how hexokinase and protein kinases exercise regulation of photosynthetic carbohydrate metabolism including the carbohydrate-mediated repression of photosynthetic gene expression is not yet clear. In addition, effects of excessive photosynthetic source capacity on the levels of H_2O_2, inorganic phosphate, pH and ATP within the chloroplasts in which central photosynthesis is performed have not been analyzed in intact plants at real times under light. A main reason seems to be the lack of appropriate methods. Therefore, further researches including those following the development of new methods are important to elucidate further the regulatory mechanism for leaf photosynthesis under excessive photosynthetic source capacity.

Recent studies using transgenic plants have shown that overexpression of Calvin cycle enzymes (sedoheptulose-1,7-bisphosphatase and fructose-1,6-bisphosphatase) or leaf plasma membrane CO_2 transport protein increases the leaf photosynthetic rate and the biomass production (Raines, 2003, 2006). Increasing plant leaf photosynthesis and thereby increasing plant matter (biomass) production seems to be an effective way to resolve the serious problems such as climatic warming and food and energy shortages. However, data obtained in the present study and those from other studies strongly suggest that excessive photosynthetic source capacity decreases the efficiency of leaf photosynthetic matter production. This means that under excessive photosynthetic source capacity, efficiency of plant matter (biomass) production decreases. There is also evidence for the excessive photosynthetic source capacity causing down regulation of photosynthesis in plants under field conditions (Okita et al., 2001; Smidansky et al., 2002, 2007). Therefore, it is strongly suggested that for the efficient improvement of plant matter (biomass) production, well-balanced improvement of source and sink would be essential. Further studies are desired for deeper and more comprehensive understanding of the regulatory mechanism of photosynthetic source-sink balance including the regulatory mechanism for leaf photosynthesis under excessive photosynthetic source capacity. Soybean plants (*Glycine max* L. Merr. cv. Tsurunoko) used in the present study from which single-rooted soybean leaves can be made are one of the important experimental materials.

5. Conclusion

Studies using single-rooted soybean leaves, each of which is constituted from a primary leaf, a short petiole and roots developed from the petiole, have implicated that there is a regulation of leaf photosynthesis through deactivation of Rubisco, which is associated with accumulation of photosynthetic carbohydrates in leaf under excessive photosynthetic source capacity. The present study using intact soybean plants from which single-rooted soybean leaves can be made has also suggested the same regulatory mechanism for leaf photosynthesis under excessive photosynthetic source capacity. It is therefore concluded that for efficient improvement of plant matter (biomass) production, well-balanced improvement of source and sink would be essential. Further studies are desired for more complete understanding of the regulatory mechanism for leaf photosynthesis under excessive photosynthetic source capacity and its application.

6. References

Anwaruzzaman, Sawada, S., Usuda, H., Yokota, A. (1995). Regulation of ribulose-1,5-bisphosphate carboxylase/oxygenase activation by inorganic phosphate through stimulating the binding of the activator CO_2 to the activation sites. *Plant & Cell Physiology*, 36: 425-433.

Baena-Gonzalez, E., Rolland, F., Thevelein, J.M., Sheen, J. (2007). A central integrator of transcription networks in plant stress and energy signaling. *Nature*, 448: 938-942.

Bhagwat, A.S. (1981). Activation of spinach ribulose-1,5-bisphosphate carboxylase by inorganic phosphate. *Plant Science Letters*, 23: 197-206.

Bradford, M.M. (1976). A rapid and sensitive method for the quantitation of microgram quantities of protein utilizing the principle of protein-dye binding. *Analytical Biochemistry*, 72: 248-254.

Bradley, F.M., Janes, H.W. (1985). Carbon partitioning in tomato leaves exposed to continuous light. *Acta Horticulturae*, 174: 293-302.

Bredmose, N.B., Nielsen, K.L. (2009). Controlled atmosphere storage at high CO_2 and low O_2 levels affects stomatal conductance and influences root formation in kalanchoe cuttings. *Scientia Horticulturae*, 122: 91-95.

Brodribb, T.J., McAdam, S.A.M. (2011). Passive origins of stomatal control in vascular plants. *Science*, 331: 582-585.

Brooks, A., Portis, A.R. (1988). Protein-bound ribulose bisphosphate correlates with deactivation of ribulose bisphosphate carboxylase in leaves. *Plant Physiology*, 87: 244-249.

Chen, J., Wang, P., Mi, H.L., Chen, G.Y., Xu, D.Q. (2010). Reversible association of ribulose-1,5-bisphosphate carboxylase/oxygenase activase with the thylakoid membrane depends upon the ATP level and pH in rice without heat stress. *The Journal of Experimental Botany*, 61: 2939-2950.

Cheng, Y., Arakawa, O., Kasai, M., Sawada, S. (2008). Analysis of reduced photosynthesis in the apple leaf under sink-limited conditions due to girdling. *Journal of the Japanese Society for Horticultural Science*, 77: 115-121.

Cheng, Y., Arakawa, O., Sawada, S. (2004). A mechanism of photosynthetic inhibition in sink-limited young apple trees by continuous light. *Horticultural Research*, 3: 393-398.

Crafts-Brandner, S.J., Egli, D.B. (1987). Sink removal and leaf senescence in soybean. *Plant Physiology*, 85: 662-666.

Crafts-Brandner, S.J., Salvucci, M.E. (2000). Rubisco activase constrains the photosynthetic potential of leaves at high temperature and CO_2. *Proceedings of the National Academy of Sciences U.S.A.*, 97: 13430-13435.

Cushman, K.E., Tibbitts, T.W., Sharkey, T.D. (1995). Constant-light injury of potato: temporal and spatial patterns of carbon dioxide assimilation, starch content, chloroplast integrity, and necrotic lesions. *Journal of the American Society for Horticultural Science*, 120: 1032-1040.

Dai, N., Schaffer, A., Petreikov, M., Shahak, Y., Giller, Y., Ratner, K., Levine, A., Granot, D. (1999). Overexpression of *Arabidopsis* hexokinase in tomato plants inhibits growth, reduces photosynthesis, and induces rapid senescence. *The Plant Cell*, 11: 1253-1266.

Globig, S., Rosen, I., Janes, H.W. (1997). Continuous light effects on photosynthesis and carbon metabolism in tomato. *Acta Horticulturae*, 418: 141-151.

Jang, J.C., Leon, P., Zhou, L., Sheen, J. (1997). Hexokinase as a sugar sensor in higher plants. *The Plant Cell*, 9: 5-19.

Jordan, D.B., Chollet, R. (1983). Inhibition of ribulose bisphosphate carboxylase by substrate ribulose-1,5-bisphosphate. *The Journal of Biological Chemistry*, 258: 13752-13758.

Kasai, M. (2008). Regulation of leaf photosynthetic rate correlating with leaf carbohydrate status and activation state of Rubisco under a variety of photosynthetic source/sink balances. *Physiologia Plantarum*, 134: 216-226.

Mackinney, G. (1941). Absorption of light by chlorophyll solutions. *The Journal of Biological Chemistry*, 140: 315-322.

Makino, A., Mae, T., Ohira, K. (1986). Colorimetric measurement of protein stained with Coomassie Brilliant Blue R on sodium dodecyl sulfate-polyacrylamide gel electrophoresis by eluting with formamide. *Agricultural and Biological Chemistry*, 50: 1911-1912.

Martin, T., Oswald, O., Graham, I.A. (2002). *Arabidopsis* seedlings growth, storage mobilization, and photosynthetic gene expression are regulated by carbon: nitrogen availability. *Plant Physiology*, 128: 472-481.

McCurry, S.D., Pierce, J., Tolbert, N.E., Orme-Johnson, W.H. (1981). On the mechanism of effector-mediated activation of ribulose bisphosphate carboxylase/oxygenase. *The Journal of Biological Chemistry*, 256: 6623-6628.

Mondal, M.H., Brun, W.A., Brenner, M.L. (1978). Effects of sink removal on photosynthesis and senescence in leaves of soybean (*Glycine max* L.) plants. *Plant Physiology*, 61: 394-397.

Moore, B., Zhou, L., Rolland, F., Hall, Q., Cheng, W.H., Lui, Y.X., Hwang, J., Jones, T., Sheen, J. (2003). Role of the *Arabidopsis* glucose sensor HXK1 in nutrient, light, and hormonal signaling. *Science*, 300: 332-336.

Murage, E.N., Watashiro, N., Masuda, M. (1996). Leaf chlorosis and carbon metabolism of eggplant in response to continuous light and carbon dioxide. *Scientia Horticulturae*, 67: 27-37.

Murage, E.N., Watashiro, N., Masuda, M. (1997). Influence of light quality, PPFD and temperature on leaf chlorosis of eggplants grown under continuous illumination. *Scientia Horticulturae*, 68: 73-82.

Okita, T.W., Sun, J., Sakulringharoj, C., Choi, S.B., Edwards, G.E., Kato, C., Ito, H., Matsui, H. (2001). Increasing rice productivity and yield by manipulation of starch synthesis. *Novartis Found Symp*, 236: 135-146 (discussion 147-152).

Paul, M.J., Foyer, C.H. (2001). Sink regulation of photosynthesis. *The Journal of Experimental Botany*, 52: 1383-1400.

Paul, M.J., Pellny, T.K. (2003). Carbon metabolite feedback regulation of leaf photosynthesis and development. *The Journal of Experimental Botany*, 54: 539-547.

Raines, C.A. (2003). Minireview The calvin cycle revisited. *Photosynthesis Research*, 75: 1-10.

Raines, C.A. (2006). Transgenic approaches to manipulate the environmental responses of the C_3 carbon fixation cycle. *Plant Cell & Environment*, 29: 331-339.

Raines, C.A. (2011). Increasing photosynthetic carbon assimilation in C_3 plants to improve crop yield: Current and future strategies. *Plant Physiology*, 155: 36-42.

Rowell, T., Mortley, D.G., Loretan, P.A., Bonsi, C.K., Hill, W.A. (1999). Continuous daily light period and temperature influence peanut yield in nutrient film technique. *Crop Science*, 39: 1111-1114.

Sage, R.F., Sharkey, T.D., Seemann, J.R. (1989). Acclimation of photosynthesis to elevated CO_2 in five C_3 species. *Plant Physiology*, 89: 590-596.

Sawada, S., Arakawa, O., Muraki, I., Echigo, H., Miyashita, M., Iwafune, M., Kasai, M. (1999). Photosynthesis with single-rooted *Amaranthus* leaves. I. Changes in the activities of ribulose-1,5-bisphosphate carboxylase and phosphoenolpyruvate carboxylase and the amounts of intermediates in photosynthetic metabolism in response to changes in the source-sink balance. *Plant & Cell Physiology*, 40: 1143-1161.

Sawada, S., Hasegawa, Y., Kasai, M., Sasaki, M. (1989). Photosynthetic electron transport and carbon metabolism during altered source/sink balance in single-rooted soybean leaves. *Plant & Cell Physiology*, 30: 691-698.

Sawada, S., Hayakawa, T., Fukushi, K., Kasai, M. (1986). Influence of carbohydrates on photosynthesis in single, rooted soybean leaves used as a source-sink model. *Plant & Cell Physiology*, 27: 591-600.

Sawada, S., Sato, M., Kasai, A., Yaochi, D., Kameya, Y., Matsumoto, I., Kasai, M. (2003). Analysis of the feed-forward effects of sink activity on the photosynthetic source-sink balance in single-rooted sweet potato leaves. I. Activation of RuBPcase through the development of sinks. *Plant & Cell Physiology*, 44: 190-197.

Sawada, S., Usuda, H., Hasegawa, Y., Tsukui, T. (1990). Regulation of ribulose-1,5-bisphosphate carboxylase activity in response to changes in the source/sink balance in single-rooted soybean leaves: the role of inorganic orthophosphate in activation of the enzyme. *Plant & Cell Physiology*, 31: 697-704.

Sawada, S., Usuda, H., Tsukui, T. (1992). Participation of inorganic orthophosphate in regulation of the ribulose-1,5-bisphosphate carboxylase activity in response to changes in the photosynthetic source-sink balance. *Plant & Cell Physiology*, 33: 943-949.

Setter, T.L., Brun, W.A. (1980). Stomatal closure and photosynthetic inhibition in soybean leaves induced by petiole girdling and pod removal. *Plant Physiology*, 65: 884-887.

Setter, T.L., Brun, W.A., Brenner, M.L. (1980). Effect of obstructed translocation on leaf abscisic acid, and associated stomatal closure and photosynthesis decline. *Plant Physiology*, 65: 1111-1115.

Smidansky, E.D., Clancy, M., Meyer, F.D., Lanning, S.P., Blake, N.K., Talbert, L.E., Giroux, M.J. (2002). Giroux, Enhanced ADP-glucose pyrophosphorylase activity in wheat endosperm increases seed yield. *Proceedings of the National Academy of Sciences U.S.A.*, 99: 1724-1729.

Smidansky, E.D., Meyer, F.D., Blakeslee, B., Weglarz, T.E., Greene, T.W., Giroux, M.J. (2007). Expression of a modified ADP-glucose pyrophosphorylase large subunit in wheat seeds stimulates photosynthesis and carbon metabolism. *Planta*, 225: 965-976.

Stessman, D., Miller, A., Spalding, M., Rodermel, S. (2002). Regulation of photosynthesis during *Arabidopsis* leaf development in continuous light. *Photosynthesis Research*, 72: 27-37.

Tibbitts, T.W., Bennett, S.M., Cao, W. (1990). Control of continuous irradiation injury on potatoes with daily temperature cycling. *Plant Physiology*, 93: 409-411.

von Caemmerer, S., Evans, J.R. (2010). Enhancing C_3 photosynthesis. *Plant Physiology*, 154: 589-592.

Ward, J.K., Tissue, D.T., Thomas, R.B., Strain, B.R. (1999). Comparative responses of model C_3 and C_4 plants to drought in low and elevated CO_2. *Global Change Biology*, 5: 857-867.

Wheeler, G.M., Tibbitts, T.W. (1986). Growth and tuberization of potato (*Solanum tuberosum* L.) under continuous light. *Plant Physiology*, 80: 801-804.

Wittenbach, V.A. (1982). Effect of pod removal on leaf senescence in soybeans. *Plant Physiology*, 70: 1544-1548.

Wittenbach, V.A. (1983). Effect of pod removal on leaf photosynthesis and soluble protein composition of field-grown soybeans. *Plant Physiology*, 73: 121-124.

Xu, D.Q., Gifford, R.M., Chow, W.S. (1994). Photosynthetic acclimation in pea and soybean to high atmospheric CO_2 partial pressure. *Plant Physiology*, 106: 661-671.

Zhou, Y.H., Yu, J.Q., Mao, W.H., Huang, L.F., Song, X.S., Nogues, S. (2006). Genotypic variation of Rubisco expression, photosynthetic electron flow and antioxidant metabolism in the chloroplast of chilled-exposed cucumber plants. *Plant & Cell Physiology*, 47: 192-199.

8

Up-Regulation of Heme Oxygenase by Nitric Oxide and Effect of Carbon Monoxide on Soybean Plants Subjected to Salinity

Guillermo Noriega, Carla Zilli, Diego Santa Cruz, Ethel Caggiano,
Manuel López Lecube, María Tomaro and Karina Balestrasse
University of Buenos Aires, Consejo Nacional de Investigaciones Científicas y Técnicas
Argentina

1. Introduction

Reactive oxygen species (ROS) are generated in small amounts in the normal metabolism of the cells and in increased amounts under many conditions of altered cell physiology; they are responsible for many kinds of cell injuries (Sies 1993) and have been shown to induce a significant reprogramming of gene expression (Colburn 1992).

Salt stress is one of the most important abiotic stresses that adversely affects soybean growth and causes significant crop loss worldwide. Salinity has always been considered a serious constraint on agricultural productivity (Hay & Porter 2006) and affects plant's physiology. Salt stress is a complex phenomenon that involves morphological and developmental changes. Two major components have been identified in this insult, osmotic stress and ion toxicity (Darwish et al. 2009). Higher plants have multiple protective mechanisms against salt stress including ion homeostasis, osmolyte biosynthesis, ROS scavenging, water transport, and transducers of long-distance response coordination. It is generally accepted that many stresses, including salinity, induce an overproduction of ROS, such as H_2O_2, O_2 •-, and $HO\cdot$, and these species are thought to be responsible for the oxidative damage associated with plant stress (Zilli et al. 2009). To counteract the toxicity of ROS, defense systems that scavenge cellular ROS have been developed in plants to cope with oxidative stress via the non-enzymatic and enzymatic systems (Demiral & Turkan 2005; Mandhania et al. 2006)

Nitric oxide (NO) acts as a signaling molecule and mediates multiple physiological processes in plants (Leitner et al. 2009). In addition, it has been implicated in responses to biotic and abiotic stresses, such as disease resistance, salinity , drought, heat stress, among others (Beligni & Lamattina 1999; Romero-Puerta et al. 2004; Corpas et al. 2009). There are several sources of NO in plants, but mainly it can be enzymatically produced by nitrate reductase and nitric oxide synthase-like enzymes (Wilson et al. 2008 and Corpas et al. 2009). NO is a reactive nitrogen species and, depending on its concentration, it produces either protective or toxic effects. A low dose of NO modulates superoxide anion formation and inhibits lipid peroxidation, resulting in an antioxidant function during stress (Boveris et al. 2000 and Santa Cruz et al. 2010). Moreover, microarray studies have shown that NO induces a large number of genes at transcriptional level, among them those of antioxidant enzymes (Parani et al. 2004). It has also been reported that Nitric oxide gives rise to signaling pathways mediating responses of specific genes to ultraviolet-B (UV-B) radiation, such as chalcone synthase and phenylalanine ammonia lyase (Mackerness et al. 2001). However, information about the role that NO plays in regulation of antioxidant enzymes to counteract salt-induced oxidative stress is rather limited.

Nitric oxide is believed to act as a signal molecule mediating responses to both biotic and abiotic stresses in plants (reviewed in Xuan et al. 2010 and Nürnberger & Scheel 2001) and its presence has been shown to induce seed germination (Liu et al. 2010), to affect growth and development of plant tissue (Beligni & Lamatina 2001, to increase iron homeostasis (Martin et al. 2009), to regulate plant maturation and senescence (Yaacov et al. 1998 and Jasid et al. 2009) to mediate abscisic acid-induced stomatal closing (Garcia-Mata & Lamattina, 2007). Recently, a few studies suggested that NO can play a role in protecting plants from oxidative stresses (Shantel et al. 2008) and NO-donor treatment protected plants from damage by increasing the activity of antioxidative enzymes.

Heme oxygenase catalyzes the oxidative degradation of heme and has well-known antioxidant properties in mammals by mean of its products biliverdin IXα and carbon monoxide (CO) (Kikuchi et al. 2005). One of the three known mammalian isoforms, heme oxygenase-1 (HO-1), is induced in animal tissues by many factors including its own substrate heme, heavy metals, UV-A radiation among others (Tomaro & Batlle 2002). While earlier studies pointed to plant HO as a source of phytochrome chromophore (Terry et al. 2002), more recent works showed that HO synthesis increases in soybean plants subjected to oxidative stress conferring resistance to a subsequent insult (Noriega et al. 2004; Balestrasse et al. 2005). Moreover, we have recently demonstrated that ROS are involved in HO-1 up-regulation in soybean leaves subjected to UV-B radiation (Yannarelli et al. 2006 and Santa Cruz et al. 2010). We hypothesized that NO may also participate in this process, as it regulates the oxidative status and mediates other UV-B responses.

The aim of the present study was to investigate whether NO or CO could protect soybean against salt-induced oxidative stress through the modulation of HO activity. Soybean plants were subjected to salt stress after pre-treatments with different concentrations of sodium nitroprussiate (SNP), a well-characterized NO-donor or CO. Overall, our results indicate that in soybean plants NO is involved in the signaling pathway leading to HO-1 up-regulation under salinity, and that a balance between NO and ROS is important to trigger the antioxidant response against oxidative stress. On the other hand pretreatment with CO did not provoke any change.

2. Materials and methods

2.1 Plant material and treatments

Surface sterilized soybean seeds (*Glycine max.* L.) (A6445RG) were germinated for 10 days in plastic pots containing vermiculite in controlled environmental chambers, with a photoperiod of 16 h, photon flux density of 175 μmol m^{-2} s^{-1}, and a day/night regime of 25/20°C. Afterwards, they were pretreated hydroponically with different sodium nitroprusiate concentrations (250-750 μM) for 72 h and then with NaCl (200 mM) for 48 h.

Carbon monoxide was generated from H_2SO_4 and formic acid (HCOOH). Stock solution was prepared by bubbling CO in a Hoagland solution for 40 min and was immediately diluted (50%) to perform analysis.

Plants were then harvested. When the effect of Zn-protoporphyrin IX (ZnPPIX) was investigated, roots were pretreated with 22 μM ZnPPIX during 4 h before addition of NaCl.

Controls were incubated in buffer. For fresh weight determination, plants were filtered, washed three times with distilled water, kept on filter paper for a few minutes to remove of excess liquid and weighed. Three different experiments were performed, with three replicated measurements for each parameter assayed

2.2 Thiobarbituric acid reactive substances (TBARS) determination
Lipid peroxidation was measured as the amount of TBARS determined by the thiobarbituric acid (TBA) reaction as described by Heath and Packer (1968). Fresh control and treated roots (0.3 g) were homogenized in 3ml of 20% (w/v) trichloroacetic acid (TCA). The homogenate was centrifuged at 3,500 x g for 20 min. To 1 ml of the aliquot of the supernatant, 1ml of 20% TCA containing 0.5% (w/v) TBA and 100 ml 4% butylated hydroxytoluene (BHT) in ethanol were added. The mixture was heated at 95°C for 30min and then quickly cooled on ice. The contents were centrifuged at 10,000 x g for 15 min and the absorbance was measured at 532 nm. Value for non-specific absorption at 600 nm was substracted. The concentration of TBARS was calculated using an extinction coefficient of 155 mM^{-1}cm^{-1}

2.3 Heme oxygenase preparation and assay
Roots (0.3 g) were homogenized in a Potter-Elvehejm homogenizer using 4 vol. of ice-cold 0.25M sucrose solution containing 1mM phenylmethyl sulfonyl fluoride, 0.2mM EDTA and 50mM potassium phosphate buffer (pH 7.4). Homogenates were centrifuged at 20,000 x g for 20min and supernatant fractions were used for activity determination. Heme oxygenase activity was determined as previously described with minor modifications (Muramoto et al. 2002). The standard incubation mixture in a final volume of 500ml contained 10mmol potassium phosphate buffer (pH 7.4), 60 nmol NADPH, 250ml HO (0.5mg protein), and 200 nmol hemin. Incubations were carried out at 378°C during 60min. Activity was determined by measuring biliverdin formation, which was calculated using the absorbance change at 650 nm employing an 1 value of 6.25mM^{-1}cm^{-1} (vis$_{max}$ 650 nm)

2.4 Glutathione determination
Non-protein thiols were extracted by homogenizing 0.3 g of roots in 3.0 ml of 0.1 N HCl (pH 2.0), and 1 g PVP. After centrifugation at 10,000 x g for 30 min at 4°C, the supernatants were used for analysis. Total glutathione (GSH plus GSSG) was determined in the homogenates spectrophotometrically at 412 nm, after precipitation with 0.1 N HCl, using yeast-glutathione reductase, 5,5' dithio-bis-(2-nitrobenzoic acid) (DTNB) and NADPH. GSSG was determined by the same method in the presence of 2-vinylpyridine and GSH content was calculated from the difference between total glutathione and GSSG (Anderson, 1985).

2.5 Classical antioxidant enzymes
Extracts for determination of catalase (CAT), ascorbate peroxidase (APX) and glutathione reductase (GR) activities were prepared from 0.3 g of roots homogenized under ice-cold conditions in 3 ml of extraction buffer, containing 50 mM phosphate buffer (pH 7.4), 1 mM EDTA, 1 g PVP, and 0.5% (v/v) Triton X-100 at 4 °C. The homogenates were centrifuged at 10,000 × g for 20 min and the supernatant fraction was used for the assays.
CAT activity was determined in the homogenates by measuring the decrease in absorption at 240 nm in a reaction medium containing 50 mM potassium phosphate buffer (pH 7.2) and 2 mM H$_2$O$_2$. The pseudo-first order reaction constant (k' = k[CAT]) of the decrease in H$_2$O$_2$

absorption was determined and the catalase content in pmol mg−1 protein was calculated using $k = 4.7 \times 10^7 \, M^{-1}s^{-1}$.

APX activity was measured immediately in fresh extracts and was assayed as described by Nakano and Asada (1981), using a reaction mixture (1 ml) containing 50 mM K-phosphate buffer (pH 7.0), 0.1 mM H_2O_2, 0.5 mM Na-Ascorbate and 0.1 mM EDTA. The hydrogen peroxide-dependent oxidation of Ascorbate was followed by a decrease in the absorbance at 290 nm (ε: 2.8 mM $^{-1}$ cm^{-1}). One unit of APX forms 1 µmol of ascorbate oxidized per minute under the assay conditions.

GR activity was measured by following the decrease in absorbance at 340 nm due to NADPH oxidation. The reaction mixture contained tissue extract, 1 mM EDTA,0.5 mM GSSG, 0.15 mM NADPH and 50 mM Tris–HCl buffer (pH 7.5) and 3 mM $MgCl_2$ (Schaedle and Bassham 1977).

2.6 Histochemical analysis

In order to analyze H_2O_2 generation roots were excised and immersed in a 1% solution of 3,3'-Diaminobenzidine (DAB) in Tris-HCl buffer (pH 6.5), vacuum-infiltrated for 5 min and then incubated at room temperature for 16 h in the absence of light. Roots were illuminated until appearance of brown colors characteristic of the reaction of DAB with H_2O_2.

In the same way to show $O_2{}^-$ production roots were excised and immersed in a 0.1% solution of NBT in K-phosphate buffer (pH 6.4), containing 10 mM Na-azide, and were vacuum-infiltrated for 5 min and illuminated until appearance of dark spots, characteristic of blue formazan precipitate.

2.7 Isolation of RNA and RT-PCR analysis

Total RNA was extracted from soybean roots by using the Trizol reagent (Gibco BRL). Four micrograms of total RNA were treated with RNase-free DNase I (Promega, CA, USA) and then 1.0 µg was reversed transcribed into cDNA using random hexamers and M-MLV Superscript II RT (Invitrogen, CA, USA). PCR reactions were carried out using *Glycine max* HO-1 and 18S specific primers, as previously described (Yannarelli and others, 2006). The PCR profile was set at 94°C for 1 min and then 29 cycles at 94°C for 0.5 min, 54°C for 1 min, and 72°C for 1 min, with a final extension at 72°C for 7 min. Each primer set was amplified using an optimized number of PCR cycles to ensure the linearity requirement for semi-quantitative RT-PCR analysis. The amplified transcripts were visualized on 1.5% agarose gels with the use of ethidium bromide. Gels were then scanned (Fotodyne Incorporated, WI, USA) and analyzed using Gel-Pro Analyzer 3.1 software (Media Cybernetics, MD, USA).

2.8 Protein determination

Protein concentration was evaluated by the method of Bradford (1976), using bovine serum albumin as a standard.

2.9 Statistics

Values in the text, figures and tables indicate mean values ± SEM. Differences among treatments were analyzed by one-way ANOVA, taking p<0,05 as significant according to Tukey's multiple range test.

3. Results

3.1 Growth parameter

Experiments were carried out in the presence of different SNP concentrations ranging from 200 to 750 µM. Root length was measured as a parameter to asses the optimal condition. Figure 1 shows that 250 µM SNP brought about a 45% increase in root length, whereas a diminution was observed under the other concentrations. Depending on its dose, NO can promote or inhibit root growth. According to these result, 250 µM SNP was chosen as the concentration to be used in pretreatment.

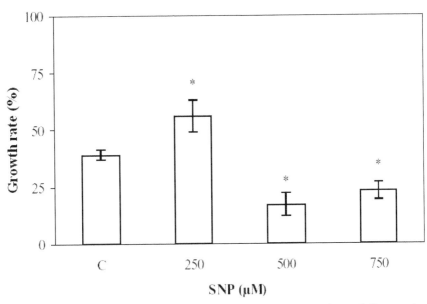

Fig. 1. Effect of different SNP concentrations on root growth. * Significant difference (p< 0.05) according to Tukey's test.

3.2 Lipid peroxidation

Increment in TBARS is a good reflection of oxidative damage to membrane lipids and other vital molecules such as proteins, DNA and RNA. Figure 2 shows that TBARS levels increased 75% respect to controls under salt treatment which is in agreement with results of other studies (Deng et al. 2010).

To complete this analysis, the effect of SNP pre-treatment was evaluated. Figure 2 indicates that in this case, membrane damage was more moderated, as indicated by a 14% augmentation respect to controls. Treatment with SNP alone did not show any difference respect to controls.

3.3 Glutathione content

GSH is a leading substrate for enzymatic antioxidant functions and it is also a known radical scavenger. Previous reports from our laboratory demonstrated that oxidative stress induces the formation of oxidant species and therefore affects GSH content in soybean plants

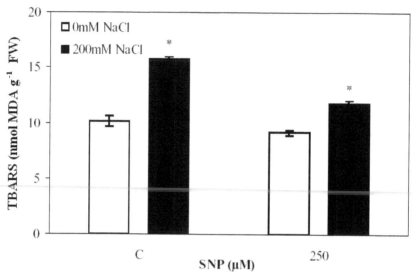

Fig. 2. Effect of salinity on TBARS formation and SNP regulation. * Significant difference (p<0.05) according to Tukey test.

(Balestrasse et al. 2001 and Noriega et al. 2004). Surprisingly, data in Figure 3 show that GSH concentration in soybean roots treated with NaCl was enhanced 3.5-fold respect to controls. Pre-treatment with SNP brought about a 4-fold augmentation respect to controls. Moreover, SNP alone provoked a 2-fold increase respect to controls.

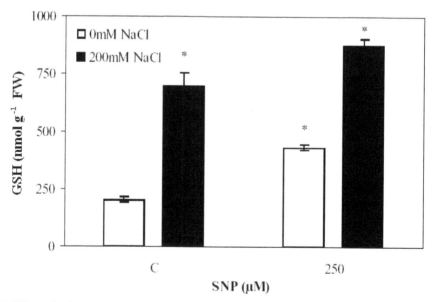

Fig. 3. Effect of salinity on GSH levels and SNP regulation. * Significant difference (p<0.05) according to Tukey test.

3.4 H$_2$O$_2$ and O$_2^-$ localization *in situ*

Accumulation of H$_2$O$_2$ and O$_2^-$ were also evaluated *in situ* by histochemical methods as shown in Figure 4a NaCl produced 32% H$_2$O$_2$ spots area versus total root area, while pretreatment with 250 µM SNP prevented this effect and spot area was similar to controls (Figure 4a). Data in Figure 4b showed that roots treated with NaCl produced 41% O$_2^-$ spots area versus total root area. Pretreatments with 250 µM SNP completely prevented the O$_2^-$ production induced by NaCl.

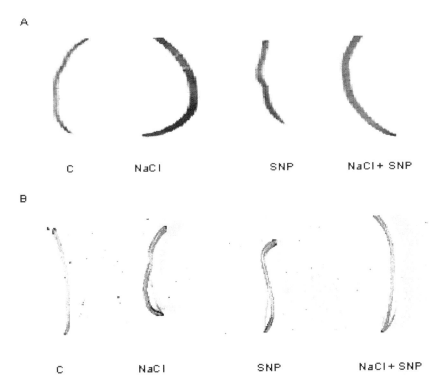

Fig. 4. Histochemical detection of H$_2$O$_2$ (A) and O$_2^-$ (B) in soybean roots. Experiments were performed as described in Materials and Methods. Pictures are representative of three different experiments with three replicated measurements for each treatment.

3.5 Effect of NO on antioxidant enzyme activities

We also investigated whether NO can modulate the activities of classical antioxidant enzymes such as CAT and APX. These are the main H$_2$O$_2$-scavenging enzymes that control ROS-mediated responses under biotic and abiotic stresses (Mittler 2002). CAT and APX activities were significantly affected by NaCl (Table 1). They were increased by 47% and 33% in NaCl-treated plants compared to controls, respectively. Moreover, CAT activity significantly augmented up to 24% with respect to controls of SNP-treated plants, whereas APX only showed a mild increase (19%). Heme oxygenase behavior was similar to that found for CAT (Table 1).

Treatment	HO-1 (U/mg protein)	CAT (pmol/mg protein)	APX (U/mg protein)
Control	0.065 ± 0.001[a]	120 ± 12[a]	0.0040 ± 0.0010[a]
NaCl	0.073 ± 0.001[b]	176 ± 9[b]	0.0053± 0.0012[a]
SNP	0.079 ± 0.002[c]	149± 2[c]	0.0047 ± 0.0010[a]
SNP+ NaCl	0.083 ± 0.004[c]	138 ± 14[c]	0.0050 ±0.0010[a]

Table 1. Antioxidant enzyme activities in soybean roots subjected to 200mM NaCl and 250 µM SNP pretreatment. Enzymatic activities were assayed as described in Materials and Methods. Different letters within columns indicate significant differences ($P < 0.05$) according to Tukey's multiple range test.

3.6 Heme oxygenase-1 activity and gene expression

Previous findings from our group demonstrated the protective role that HO-1 plays against oxidative stress in soybean plants (Noriega et al. 2004 and Balestrasse et al. 2005). Figure 5 indicates that salt stress caused HO-1 mRNA induction (13%, respect to controls). This enhancement is positively correlated with enzyme activity (Table 1). Pretreatment with 250 µM SNP brought about an augmentation of gene expression in control plants (21%), as well as salt treated plants (27%) (Figure 5). Once again, this behavior was also found when enzyme activity was determinated (Table 1) .These results indicate on one hand, that NO

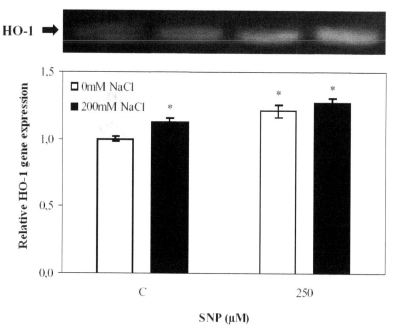

Fig. 5. HO-1 mRNA expression was analyzed by semi-quantitative RT-PCR as described in Materials and Methods. The 18S amplification band is shown to confirm equal loading of RNA and RT efficiency. Relative HO-1 transcript expression taking control as 1 U. Data are means of three independent experiments and bars indicate SE. *Significant differences (P < 0.05 according to Tukey test).

induces HO-1 more efficiently than NaCl, and on the other hand, both compounds have a synergic effect on this induction. To asses whether HO-1 is involved in the protection against NaCl exerted by NO, experiments were carried out in plants treated with ZnPPIX, a well known irreversible HO-1 inhibitor. Plants with inhibited HO-1 activity can not cope with NaCl insult (data not shown). We can assume that protection exerted by SNP may be due to the augmentation of the activity of this antioxidant enzyme.

3.7 Effect of NO and CO
3.7.1 Glutathione content
As already stated, there is a positive relationship between NO content and GSH levels (Figure 3). This result prompted us to investigate whether HO is involved in the regulation of this tripeptide. To fulfill this purpose, experiments were carried out in plants treated with ZnPPIX and then subjected to NO (HO inductor) or CO (HO reaction product) for 48 h before salt stress. Afterwards, GSH content (Figure 6) as well as HO-1 gene expression (Figure 8) was determinated.

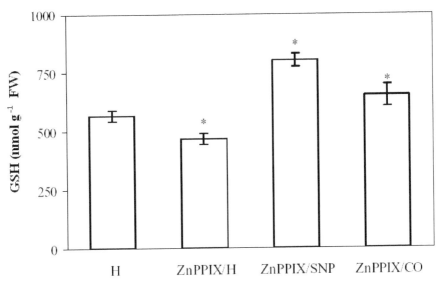

Fig. 6. Effect of NO and CO on GSH content. (H) Control plants, (ZnPPIX/H) plants pretreated with ZnPPIX and then with H; (ZnPPIX/SNP) plants pretreated with ZnPPIX and then with SNP; (ZnPPIX/CO) plants pretreated with ZnPPIX and then with CO as described in Materias and Methods. * Significant difference (p<0.05) according to Tukey's test.

In plants pretreated with ZnPPIX for 72 h before Hoagland (H) treatment (ZnPPIX/H), GSH level diminished 20% respect to controls (H). Figure 6 shows that NO (ZnPPIX/SNP) as well as CO (ZnPPIX/H) enhanced GSH levels (40% and 15%, respectively).

3.7.2 Glutathione reductase activity
Taking into account the fact that GSH synthesis is affected by HO-1 inhibition and NO pretreatment GR activity was determinated under the same conditions. Figure 7 indicates a

positive relationship between GSH levels and GR activity. Enzyme activity (GR) diminished 22% respect to controls when HO was inhibited, but an increase was detected in plants treated with NO and CO (33% and 26%, respectively).

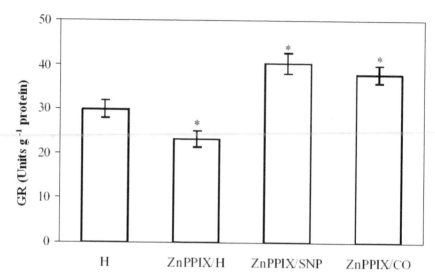

Fig. 7. Effect of NO or CO on GR activity. (H) Control plants, (ZnPPIX/H) plants pretreated with ZnPPIX and then with H; (ZnPPIX/SNP) plants pretreated with ZnPPIX and then with SNP; (ZnPPIX/CO) plants pretreated with ZnPPIX and then with CO as described in Materias and Methods. * Significant difference (p<0.05) according to Tukey's test.

3.7.3 HO-1 gene expression
Figure 8 shows HO-1 gene expression under different conditions. ZnPPIX/SNP treatment brought about a 20% augmentation respect to controls. This increase is positively correlated with GSH content and GR activity. On the other hand, CO did not show any effect. It is interesting to note that the enhancement of GSH content is not related to oxidative stress, since TBARS levels in roots of SNP and CO treated plants do not differ from controls. In contrast, HO inhibition brought about an enhancement (28%) in TBARS levels.

4. Discussion

In a previous work, we found that SNP pre-treatment ameliorates Cd-induced oxidative stress and modulates HO-1 gene expression in soybean plants (Noriega et al. 2007). Taking into account the fact that NO is involved in various signaling pathways, in the present study we evaluated whether this molecule could enhance HO activity conferring a major protection against salt stress.

Our data demonstrated that, depending on its concentration, NO can improve the plant antioxidant response against salinity. This model was appropriate to determine the beneficial effect of exogenously added NO. While the lower dose of SNP did not reduce the oxidative damage (data not shown), the application of 500 or 750μM SNP showed a deleterious effect suggesting a pro-oxidant behavior of NO at these concentrations (Figure 1).

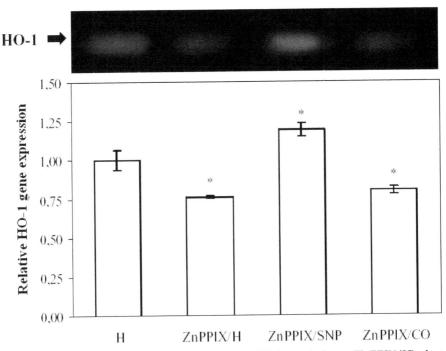

Fig. 8. Effect of NO or CO on HO-1 gene expression. (H) Control plants, (ZnPPIX/H) plants pretreated with ZnPPIX and then with H; (ZnPPIX/SNP) plants pretreated with ZnPPIX and then with SNP; (ZnPPIX/CO) plants pretreated with ZnPPIX and then with CO as described in Materias and Methods. * Significant difference (p<0.05) according to Tukey's test.

The pre-treatments with 250 µM SNP effectively ameliorated NaCl-induced oxidative stress, as indicated by the decrease in H_2O_2 and O_2^- formation (Figure 4), preventing TBARS formation (Figure 2) and enhancing GSH content (Figure 3). The activities of classical antioxidant enzymes, such as CAT and APX were also augmented by SNP treatment, instead of the drastically diminution observed with salinity alone (Table 1). These data are in agreement with reports showing a protective effect of NO in plants subjected to other stresses (Zhao et al. 2004; Shi et al. 2005 and Noriega et al. 2007). Nevertheless, the molecular mechanism that mediates NO enhancement of antioxidant enzyme activities is not completely understood. Interestingly, we found that HO and CAT activities had a similar behaviour with respect to SNP pre-treatment under salinity (Table 1). A recent study showed that the time-course of induction of those enzymes in soybean nodules subjected to Cd stress is related (Balestrasse et al. 2008). These results suggest a close relationship between the signal transduction pathways involved in the response of HO and CAT after oxidative stress generation and support the antioxidant role of HO. In addition, there was a positive correlation between HO-1 transcript levels and enzyme activity (Figure 5 and Table 1). Previous reports have also demonstrated that the enhancement of HO activity is associated with an increase in HO-1 transcript levels and protein content (Yannarelli et al. 2006 and Balestrasse et al. 2008). Although this mechanism can account for the changes observed in HO activity, the incidence of post-translational modifications or different HO isoforms under stress conditions needs to be addressed. Experiments carried out in plants treated with SNP in the absence of NaCl showed that NO itself can up-regulate HO-1 mRNA expression, but to a lesser extent (Figure 5). This observation indicates that a certain balance between NO and ROS is required to trigger the full response. Interestingly, a recent report found that the ROS–NO ratio is important to elicit ROS-activated stress responses and cell death regulation in plant leaves during ozone exposure (Ahlfors et al. 2009). Moreover, new evidence suggests that plastids and peroxisomes are important regulators of NO levels in plants (Corpas et al. 2009 and Gas et al. 2009).

Biliverdin, one of the products of the HO, is an efficient scavenger of ROS and it can account for the antioxidant properties of this enzyme both in animals and plants (Otterbein et al. 2003 and Noriega et al. 2004). More recently, it has been shown that CO released by HO is an important signal molecule for the tolerance mechanisms against cadmium and salt stress (Han et al. 2008). It would be interesting to determine whether CO could also play a role in the defense against salinity in soybean plants.

Pretreatment with ZnPPIX decreased HO-1 expression (Figure 8) and increased parameters of oxidative stress. When the inhibitor was added before NO or CO treatment, HO-1 expression as well as GSH content (Figure 6) and GR activity were increased (Figure 7). These results let us suppose that a close relationship between HO-1 induction and GSH content could exist. Taking together, these data provide evidence of one of the possible roles that NO, as well as CO could play against oxidative insult.

5. Conclusion

The present study together with previous results (Balestrasse et al. 2008 and Zilli et al. 2008) support the protective role of HO in soybean plants against salinity. Data here reported let us understand the mechanisms involved in HO response in NaCl-treated soybean plants. This model proposes that NO is implicated in the HO signaling pathway and, together with ROS, modulates the activity of this enzyme under salinity. In plants treated with ZnPPIX, CO did not induce HO-1, but an augmentation of GSH levels as well as GR activity was observed. On the other hand, NO not only caused a more important enhancement in GSH content and GR activity, but also brought about the induction of HO-1. Moreover, NO can enhance the antioxidant system allowing an improved plant defense to the subsequent oxidative insult. Interestingly, while NO may directly potentiate NaCl-induced HO-1 transcription, pre-treatment with SNP followed by salinity stress may protect and enhance by inducing free radical scavenging enzymes and GSH. An appropriate balance of ROS–NO is necessary to trigger the full HO response. In contrast to other stress conditions, induction of HO-1 occurs together with an enhancement of GSH levels and GR activity. In conclusion, the present study provides new insights into the molecular response of soybean plants to salinity and also evidences that HO plays an important role during stress conditions.

6. Acknowledgments

We thank Dr T. Kohchi for kindly providing the Arabidopsis HO-1 antibodies. This work was supported by grants from the Universidad de Buenos Aires (Argentina) and from Consejo Nacional de Investigaciones Científicas y Técnicas (CONICET) (Argentina). M.L.T and K.B.B. are career investigators from CONICET.

7. References

Ahlfors, R.; Broschél, M.; Kollist, H.; Kangasjärvil, J. (2009). Nitric oxide modulates ozone-induced cell death, hormone biosynthesis and gene expression in Arabidopsis thaliana. *Plant Journal*, Vol.58, No.1, pp.1-12, ISSN 0960-7412.

Anderson, E. (1985). Determination of glutathione and glutathione disulfide in biological samples, *Methods in Enzymology*. Vol.113, pp. 548-554, ISSN 0076-6879

Balestrasse, K.; Gardey, L.; Gallego, S. & Tomaro, M. (2001). Response of antioxidative defense system in soybean nodules and roots subjected to cadmium stress. *Australian Journal of Plant Physiology*, Vol.28, pp.497–504. ISSN 0310-784.

Balestrasse, K.; Noriega, G.; Batlle; A. & Tomaro, M. (2005). Involvement of Heme Oxygenase as Antioxidant Defense in Soybean Nodules, *Free Radical Research*, Vol.39, No.2, pp.145-51. ISSN 1071-5762.

Balestrasse, K.; Zilli, C. & Tomaro, M. (2008). Signal transduction pathways and haem oxygenase induction in soybean leaves subjected to salt stress, *Redox Report*,Vol.13; No. 6, (December 2008), pp. 255-262, ISSN 1351-0002.

Balestrasse, K.;Gardey, L.; Gallego, S. & Tomaro, M. (2001). Response of antioxidant defence system in soybean nodules and roots subjected to cadmium stress, *Australian Journal of Plant Physiology*. Vol.28, pp.497–504, ISSN 0310-7841.

Beligni, M. & Lamattina, L. (1999). Is nitric oxide toxic or protective?. *Trends in Plant Science*, Vol.4, No.8, (August 1999), pp. 299-300, ISSN.

Beligni, M. & Lamattina,L. (2001). Nitric oxide: a non-traditional regulator of plant growth, *Trends in Plant Science*, Vol6, No. 11, (November 2001), pp. 508-509, ISSN 1360-1385.

Boveris, A. & Poderoso, J. (2000). Regulation of Oxygen Metabolism by Nitric Oxide, *Nitric Oxide*, Vol.51, pp. 355-368, ISSN 1089-8306.

Bradford, M. (1976). A rapid and sensitive method for the quantitation of microgram quantities of protein utilizing the principle of protein-dye binding, *Analytical Biochemistry*, Vol.72, pp.248-254, ISSN 0003-2697.

Colburn, NH.(1992). Gene regulation by active oxygen and other stress inducers, Spatz L, Bloom AD (eds) *Biological consequences of oxidative stress*, pp 121–137, ISBN 10: 0195072960, Oxford University Press, New York.

Corpas, F.; Palma, J.; Del Río, L. & Barroso, J. (2009). Evidence supporting the existence of l-arginine-dependent nitric oxide synthase activity in plants, *New Phytologist*, Vol.184, pp.1–3, ISSN *0028-646X*.

Darwish, E.; Testerink, C.; Khalil, M.; El-Shihy, O. & Munnik, T. (2009). Phospholipid Signaling responses in salt-Stressed Rice Leaves. *Plant and Cell Physiology*, Vol.50, No.5, pp.986-99, ISSN 0032-0781.

Demiral, T. & Turkan, I. (2005). Comparative lipid peroxidation, antioxidant defense systems and proline content in roots of two rice cultivars differing in salt tolerance, *Environmental and Experimental Botany* , Vol.53, pp. 247-257, ISSN 0098-8472.

Garcia-Mata, C. & Lamattina, L. (2007). Abscisic acid (ABA) inhibits light-induced stomatal opening through calcium- and nitric oxide-mediated signaling pathways *Nitric Oxide*, Vol.17, No.3-4, (November-December 2007), pp.143-151, ISSN 1089-8306.

Harminder, P.; Shalinder Kaur, S.; Batish, D.; Sharma, V.; Sharma, N. & Kohli, R. (2009). Nitric oxide alleviates arsenic toxicity by reducing oxidative damage in the roots of *Oryza sativa* (rice), *Nitric Oxide*, Vol.20, No.4, (June 2009), pp. 289-297, ISSN 1089-8306.

Hay, R. & Porter, J. (2006). *The physiology of crop yield*, Wiley-Blackwell Publishing, ISBN 1405108592, Singapore.

Heath, R. & Packer, L. (1968). Photoperoxidation in isolated chloroplasts. I Kineties and stoichiometry of fatty acid peroxidation, *Archives of Biochemistry and Biophysics*, Vol.125, pp.189-198, ISSN 0003-9861.

Jasid, S.; Galatro, A.; Villordo, J.; Puntarulo, S. & Simontacchi, M. (2009).Role of nitric oxide in soybean cotyledon senescence, *Plant Science*, Vol. 176, No. 5, (May 2009), pp.662-668, ISSN 0168-9452.

Kikuchi, G.;Yoshida, T. & Noguchi, M. (2005). Heme oxygenase and heme degradation, *Biochemistry and Biophysical Research Communication*, Vol. 338, pp.558–567, ISSN 0006-291X.

Leitner , M.; Vandelle, E.; Gaupels, F.;Bellin,D.& Delledonne, F. (2009). NO signals in the haze Nitric oxide signalling in plant defence, *Current Opinion in Plant Biology*, Vol.12, pp.451–458. ISSN 1369-5266.

Liu, Y.; Xu, S.; Ling, T.; Xu, L. & Shen, W. (2010). Heme oxygenase/carbon monoxide system participates in regulating wheat seed germination under osmotic stress involving the nitric oxide pathway, *Journal of Plant Physiology* , Vol.167, pp. 1371–1379, ISSN 0176-1617.

Mackernessa, S.; Johna, F.; Jordan, B. & Thomasa, B. (2001). Early signaling components in ultraviolet-B responses: distinct roles for different reactive oxygen species and nitric oxide, *FEBS Letters*, Vol.489, pp.237-242, ISSN 0014-5793.

Mandhania, S.; Madan, S. & Sawhney, V. (2006). Antioxidant defense mechanism under salt stress in wheat seedlings, *Biologia Plantarum*, Vol.227, pp. 227- 231, ISSN 0006-3134.

Martin, M.; Rodríguez Colman,M.; Gómez-Casati, D.; Lamattina, L. & Julián, E. Zabaleta. (2009). Nitric oxide accumulation is required to protect against iron-mediated oxidative stress in frataxin-deficient Arabidopsis plants, *FEBS Letters*, Vol.583, No.3, (February 2009), pp. 542-548, ISSN 0014-5793.

Mittler, R. (2002) Oxidative stress, antioxidants and stress tolerance, *TRENDS in Plant Science*, Vol.7, No.9, (September 2002),pp.521-546, ISSN 1360-1385.

Muramoto, T.; Tsurui, N.; Terry, M.; Yokota, A. & Kohchi, T. (2002). Expression and biochemical properties of a ferredoxin-dependent heme oxygenase required for phytochrome chromophore synthesis. *Plant Physiology*, Vol.130, pp.1958-1966, ISSN 0032-0889.

Nakano, Y. & Asada, K. (1981). Hydrogen peroxide is scavenged by ascorbate-specific peroxidase in spinach chloroplast. *Plant and Cell Physiology*, Vol.22, pp.867-880, ISSN 0032-0781.

Noriega, G.; Balestrasse, K.; Batlle, A & Tomaro, M. (2004). Heme oxygenase exerts a protective role against oxidative stress in soybean leaves. *Biochemical and Biophysical Research Communications*. Vol.323, pp. 1003-1008, ISSN 0006-291X.

Noriega, G.; Yannarelli, G.; Balestrasse, K.; Batlle, A. & Tomaro, M. (2007). The effect of nitric oxide on heme oxygenase gene expression in soybean leaves, *Planta*, Vol. 226, pp.1155–1163, ISSN 0032-0935.

Nürnberger, T. & Scheel, D. (2001) Signal transmission in the plant immune response, *TRENDS in Plant Science*, Vol.6, No.8 (August 2001), pp. 551-579, ISSN 1360-1385.

Otterbein, L.; Soares, M.; Yamashita, K. & Bach, F. (2003). Hemeoxygenase-1:unleashing the protective properties of heme. *Trends Immunology*, Vol.24, pp.449-455, ISSN 1471-4906.

Parani, M. ; Rudrabhatla. S.; Myers, R.; Weirich, H.; Smith, B.; Leaman, B. & Goldman, S. (2004). Microarray analysis of nitric oxide responsive transcripts in *Arabidopsis*, *Plant Biotechnology Journal*, Vol.2, pp.359–366, ISSN 14677644.

Romero-Puertas, M.; Rodríguez-Serrano, M.; Corpas, F.; Gómez, M.; Del Río, L. & Sandalio, L. (2004). Cadmium-induced subcellular accumulation of $O_2 \cdot^-$ and H_2O_2 in pea leaves, *Plant and Cell Environment*, Vol.27, pp.1122–1134, ISSN 0140-7791.

Santa-Cruz, D.;Pacienza, N.; Polizio A., Balestrasse K.;Tomaro M. & Yannarelli, G. (2010). Nitric oxide synthase-like dependent NO production enhances heme oxygenase up-regulation in ultraviolet-B-irradiated soybean plants. *Phytochemistry*, Vol.71, pp.1700–1707, ISSN.

Shaedle, M. & Bassham, J. (1977). Chloroplast glutathione reductase, *Plant Physiology*, Vol. 59, pp.1011–1012, ISSN 0031-9422.

Shantel, A.; Fowler, R.; Virgen, A.; Gossett, D.; Banks, S. & Rodriguez, S. (2008). Opposing roles for superoxide and nitric oxide in the NaCl stress-induced upregulation of antioxidant enzyme activity in cotton callus tissue, *Environmental and Experimental Botany*, Vol.62, No.1, (January 2008), pp. 60-68, ISSN 0098-8472.

Shi, S.; Wang, G.; Wang, Y.; Zhang, L. & Zhang, L. (2005). Protective effect of nitric oxide against oxidative stress under ultraviolet-B radiation, *Nitric Oxide*, Vol.13, pp.1–9, ISSN 1089-8603.

Sies, H. (1993). Damage to plasmid DNA by singlet oxygen and its protection. *Mutation Research/Genetic Toxicology*, Vol. 299, No.3-4, May 1993, pp 183-191, ISSN 1383-5718.

Terry, M.; Linley, P. & Kochi, T. (2002). Making light of it: the role of plants heme oxygenases in phytochrome chromophore synthesis, *Biochemical Society Transactions*, Vol. 30, pp. 604–609, ISSN 0300-5127.

Terry, M.; Linley, P. & Kochi, T. (2002). Making light of it: the role of plants heme oxygenases in phytochrome chromophore synthesis, *Biochemical Society Transactions*, Vol. 30, pp.604–609, ISSN 0300-5127.

Tomaro, M., Batlle, A. (2002). Bilirubin: its role in cytoprotection against oxidative stress. *International Journal of Biochemistry and Cell Biology*, Vol.34, pp.216–220. ISSN 1357-2725.

Wilson, I.; Neill, S. & Hancock, J. (2008).Nitric oxide synthesis and signalling in plants, *Plant Cell and Environment*, Vol.31, pp.622–631, ISSN 0140-7791.

Xuan, Y.; Zhou, S.; Wang, L.; Cheng, Y. & Zhao L. (2010). Nitric Oxide Functions as a Signal and Acts Upstream of AtCaM3 in Thermotolerance in Arabidopsis Seedling, *Plant Physiology*. Vol.153, pp.1443-1444, ISSN 0032-0889.

Yaacov,Y.; Leshem, R.;Wills, B. & Veng-Va Ku, W. (1998). Evidence for the function of the free radical gas — nitric oxide (NO•) — as an endogenous maturation and senescence regulating factor in higher plants, *Plant Physiology and Biochemistry*, Vol. 36, No. 11, (November 1998), pp. 825-833, ISSN 0981-9428.

Yannarelli, G.; Noriega, G.; Batlle, A. & Tomaro, M. (2006). Heme oxygenase up-regulation in ultraviolet-B irradiated soybean plants involves reactive oxygen species, *Planta*, Vol.224, pp.1164–1172, ISSN 0032-0935.

Zhao, L.; Zhang, F.; Guo, J.; Yang, Y.; Li, B. & Zhang, L. (2004). Nitric oxide functions as a signal in salt resistance in the calluses from two ecotypes of reed (*Phragmites communis Trin.*). *Plant Physiology,* Vol.134, pp.849–857. ISSN 0032-0889.

Zilli, C.; Santa-Cruz, D.; Yannarelli, G.; Noriega, G.; Tomaro, M. & Balestrasse, K. (2009). Heme Oxygenase Contributes to Alleviate Salinity Damage in *Glycine max* L. Leaves, *International Journal of Cell Biology.* Published online (September 2009), PMCID: PMC2809017.

9

The Asian Soybean Rust in South America

Gustavo B. Fanaro and Anna Lucia C. H. Villavicencio
Instituto de Pesquisas Energéticas e Nucleares (IPEN)
Brazil

1. Introduction

Soybean is infected by two species of fungi that cause the rust: the *Phakopsora meibione* (Arth.) Arth. (American soybean rust), which is native from American continent, existing from Puerto Rico to southern Brazil and not cause concerns for farmers and the *Phakopsora pachyrhizi* Sydow & Sydow (Asian soybean rust), a serious disease which causes a high yield losses. The differentiation of these two species is only possible through DNA testing (Yorinori & Lazzarotto, 2004).

The *P. meibione* is the less aggressive soybean rust species and was reported in the western hemisphere, South and Central America and Caribbean. It was reported in Puerto Rico in 1913, Mexico in 1917 and Cuba in 1926 on hyacinth bean and some other leguminous species, but only in Puerto Rico in 1976 was related on soybean (Pivonia & Yang, 2004). This species occurs under mild temperatures (average below 25 °C) and high relative humidity (Yorinori & Lazzarotto, 2004).

The *P. pachyrhizi* was described as a pathogen on the legume *Pachyrhizus erosus* (L.) Urb. (well-know as jacatupé in Brazil; jícama or pois patate in France; jícama, yam and mexican turnip in English language and jícama, pipilanga, yacón or nabo mexicano in Spanish language) in Taiwan, published by Sydow & Sydow in 1914 and can infect many leguminous species in numerous orders of the family Leguminosae (Deverall et al., 1977).

The *P. pachyrhizi* was first identified in Japan in 1902, and then was detected in India (1906), Australia (1934), China (1940), in Southeast Asia (1950s) and Russia (1957). For many years it remained confined to Asia and Australia, until to be found in Hawaii in 1994 and in Africa continent (from Uganda to South Africa) in 1997 (Begenisic et al., 2004).

P. pachyrhizi was first identified in the America continent in March 2001 in Paraguay, which caused yield reduction of 1,100 kg/hectare. In May, it was also found in Paraná (Brazil). In 2001/02 harvest, the disease recurred throughout Paraguay and was also found in Argentina, Bolivia and in several states of Brazil. In the worst hit places, the reductions in grain yield were estimated between 10% and 80% (Yorinori, 2002).

The fungal inoculum, for the initial outbreak in South America, is thought to originated from southern Africa where soybean rust has been observed since the late 1990s (Scherm et al., 2009). Since 1994, the disease has been identified by several countries, damaging up to 40% of crops in Thailand, 90% in India, 50% in the south of China and 40% in Japan (Hartman et al., 1991; Mendes et al., 2009). In the United States, this disease was first reported at the Louisiana State University AgCenter Research Farm in 2004, but yield loss was not as high as those reported from other countries (Cui et al., 2010).

Soybean plants are susceptible to the fungus at all growth stages. As a general rule, the earlier a crop is attacked, the higher will be the loss (Mendes et al., 2009), however, if the attack occurs at flowering and pod filling stage, which is commonly observed in soybean fields, the yield reducing can be higher than in others stages (Kawuki et al., 2004).

2. Contamination

The Asian soybean rust is one of the most destructive diseases of soybean because it produces a high amount of airborne spores that can infect large areas of soybeans and cause significant yield loss. The fungal spores (uredospores) are deposited on leaves in the lower region of the canopy through the rain or wind transport from nearby plants during the growing season (Huber & Gillespie, 1992). The figure 1 shows a soybean leave contaminated with Asian soybean rust.

The *P. pachyrhizi* is highly moisture dependent, requiring at least 6 hours of free moisture on the lower trifoliolates to starts the contamination. Warm temperature is ideal, but not limiting, since the disease can be established between 15 °C and 30 °C (Embrapa, 2005). These moist conditions can be achieved through any form of wetness as drizzle, mist, fog or dew and the minimum duration of wetness is dependent on the ambient temperature after spore deposition (Schmitz & Grant, 2009).

Fig. 1. (Godoy et al. 2009). Soybean leaf contaminated with Asian soybean rust.

Early rust symptoms are characterized by small dots of 1 to 2 mm of diameter, darker than the healthy leaf tissue with a greenish to greenish gray coloration. On the local corresponding at the dark spot, there is initially a tiny lump, like a bubble formed by burning, showing the early formation of the fruiting structure of fungi. As soon as the death of infected tissues, the blemishes increase in size and acquire a reddish brown color. The uredospores, initially has a crystalline color, become beige and accumulate around the pores or are carried by the wind and the number of uredias per point can vary from one to six. The uredias that no longer sporulete shows the pustules with open pores, which allows distinguish them from bacterial pustule, which often causes confusion comparison (Bromfield, 1980; Embrapa, 2004).

The infection causes rapid browning and premature leaf fall, preventing the full grain formation. The earlier the defoliation, smaller is the grain size and lower is the yield and quality. In severe cases, when the disease reaches the stage of the soybean pod formation, it can cause abortion and drop of the pods, resulting in a total loss of income (Constamilan, 2002; Godoy & Canteri, 2004; Soares et al., 2004).

The life cycle is typical of the majority of other rust fungi (Fig. 2) and their uredospores are easily transported by air currents and disseminated hundreds of kilometers in few days (Tremblay et al., 2010).

Once germination occurs, the uredospore produces a single germ tube (GT) that grows across the leaf surface until it reaches an appropriate surface where an appressorium (AP) forms. This penetration occurs between 7-12h after the spore lands on the leaf adaxial surface. Appressoria form over anticlinal walls or over the center of epidermal cells, but rarely over stomata, in contrast to the habit of many other rusts. Thus, penetration is direct rather than through natural openings or through wounds in the leaf tissue. Approximately twenty hours after the spore landing, the *penetration hyphae* (PH), stemming from the appressorium cone, pass through the cuticle to emerge in the intercellular space where a septum is formed to produce the *primary infection hypha* (IH). This IH grows between palisade cells to reach the spongy mesophyll cells where it forms the haustorium (H) (Tremblay et al., 2010).

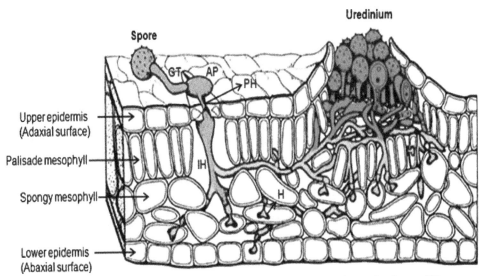

Where: GT, germ tube; AP, appressorium; PH, penetration hyphae; IH, infection hyphae and H, haustorium.

Fig. 2. (Hahn, 2000 apud Tremblay et al., 2010). Internal structure of a typical dicotyledon leaf showing the different cell layers and infection by a rust fungus.

Once this first stage has been reached, about 4 days after spore landing, additional hyphae emerge and spread through the entire spongy mesophyll layer of cells where many other haustoria are formed. At approximately 6 days after infection, some necrosis of epidermal cells occurs which is visible at the adaxial surface of the leaves (Fig. 3a). Hyphae aggregate and a uredinium arise in the spongy mesophyll cell layer. Uredinia can develop 6-8 days after spore landing and development might extend up to 4 weeks (Tremblay et al., 2010).

The first uredospores produced by the uredinium emerge at the abaxial leaf surface in 9-10 days after spore landing and spore production can be observed for up to 3 weeks. High rate of sporulation is typical of a susceptible reaction where lesions on the upper surface of the leaf are tan (Fig. 3b). Plants classified as resistant develop a dark, reddish-brown lesion with few or no spores (Fig. 3c) (Tremblay et al., 2010).

Soybean rust diagnosis is usually performed by experienced plant pathologists or plant disease diagnosticians, but nowadays, several technologies are being performed as crop health sensor, either optical or electronic or bio-electronic based to improve the perform crop disease diagnosis (Cui et al., 2010).

A useful tool that can be used as many by experienced professionals as amateurs is the diagrammatic scale to assess the severity of rust. It is very important once provides data on the severity of contamination in the plantation and the result should be informed when the competent organs are contacted, as well as help to define the goals for fungicides treatment. There are several types of scales such as developed by Godoy et al. (2006) (Fig. 4a) and Martins et al. (2004) (Fig. 4b) witch is highly recommended to be used together.

(3a)

(3b)

(3c)

Fig. 3. (Tremblay et al., 2010). Symptoms observed on soybean leaves. (a) Yellow mosaic discoloration (b) Tan lesions and (c) reddish-brown lesions.

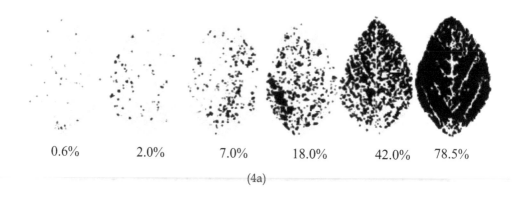

0.6%   2.0%   7.0%   18.0%   42.0% 78.5%

(4a)

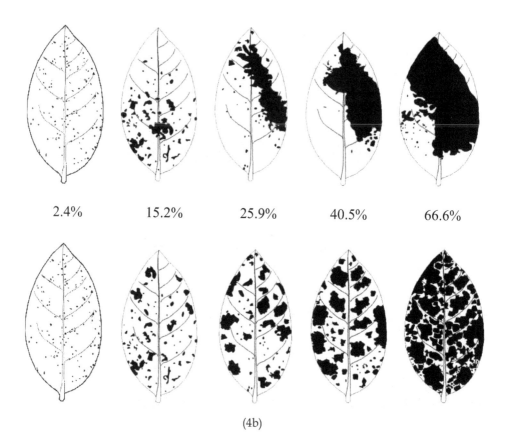

2.4%   15.2%   25.9%   40.5%   66.6%

(4b)

Fig. 4. Two diagrammatic scales of Asian soybean rust severity with percentage that represents the area of disease contamination.

3. The Asian soybean rust in South America

3.1 Argentina

The soybean crop has been converted from the middle of the nineties in the main seasonal crop of Argentina, both in area planted and in its total production (CAS, 2008). During 2004/05 season, the area devoted to soybeans was 14.4 million hectares (51% of the total planted with cereals and oilseeds), producing 38 million tones. In the season 2005/06, the soybean planting grew up to 15 and 15.3 million hectares (PNR, 2011a). In the 2006/07 season, reached a volume of 47.5 million tonnes, representing 50% of total the country grain production estimated at 94.4 million tones (CAS, 2008).

The Asian soybean rust first appeared in Argentina at the end of the 2001/02 season, in a test group in the town of Alem, province of Misiones. The infected plants samples were sent to the United States for identification by molecular analysis techniques and confirmed that the pathogen *P. pachyrhizi* was present. This finding coincided with the species identified in Brazil and Paraguay (Begenisic et al., 2004).

In the following season (2002/03), the soybean rust was detected at the end of the cycle, by a group of technicians from "Instituto Nacional de Tecnología Agropecuaria – INTA" (National Institute of Agricultural Technology) in test samples from the town of Cerro Azul, province of Misiones and two commercial lots located in the town of Gobernador Virasoro, in the province of Corrientes. Although this pathogen has penetrated the national territory, it was far from the main producing areas of the country. Because of the history of the disease, there were high producers and technicians concern about the losses that could result in the coming years (Begenisic et al., 2004).

As a result, at the beginning of the 2003/04 season, bearing in mind that rust had caused heavy losses in neighboring countries and displaying that it could become a serious concerns for Argentina, despite until that moment the rust had a little history in the country, the Ministry of Agriculture launched the "Programa Nacional de Roya de la Soja – PNRS" (National Program for Soybean Rust), coordinating activities with various agencies and public and private institutions in order to minimize the possible impact of the disease in the country (PNRS, 2011a).

The implementation of the PNRS was the first opportunity in which all public institutions have joined forces, in a cooperative manner, carrying out activities in a coordinated way and contributing in those components of the program according to their specific duties incumbent on each institution (PNRS, 2011a).

During crop season of 2004/05, the disease was detected in 13 provinces, including provinces witch the disease was detected for the first time, representing the advance of the disease significantly when compared with the last season. In most of the contaminated provinces, the rust did not cause economic losses due to its late appearance in the crop, with the exception of the province of Entre Rios where the disease was much more severe, reporting significant yield losses up to 30% (PNRS, 2011b).

These low levels of contamination can be explained for (PNRS, 2011b):

- Drought conditions and high temperatures occurring in some regions of Brazil, Paraguay and Bolivia, neighborhood to Argentina, that were not conducive to serious infections and consequently to the high seed production regional;
- The drought in Argentina during the first months of 2005 that prevented the infection conditions for generating a second "weather barrier";
- The limited survival of the fungus because the Argentina has not double cropping of soybeans in the year as in Brazil or Bolivia;

- The high level of adoption of fungicide use in Brazil, Paraguay and Bolivia.

The fungicides recommended by Argentina are the strobilurin, triazoles and their mixtures. The decision to apply is at the first signs and/or when was possible to anticipated the diagnosis in the field or when they are found in areas close to their lots and recorded favorable environmental conditions to ensure at least 7-10 hours of wet leaf and average temperatures of 22°C.

The Argentina has an official monitoring system that allow to analyze a large number of samples for the detection and disease monitoring through the website "www.sinavimo.gov.ar" (in Espanish).

3.2 Bolivia

Bolivia is the eighth country in soybeans production and is the fourth in the South America, after Brazil, Argentina and Paraguay and is one of the most important and is the successful of the national economy, due to growth in primary production, processing and export during the last fifteen years (CAS, 2008).

In Bolivia, the Asian soybean rust was first detected in the winter crop season of 2003 at Ichilo, City of Yapacani and is currently distributed throughout in all soybean crops of Santa Cruz and affecting the crop of Tarija (Yacuiba). Before the advent of Asian rust, the number of applications of fungicides for control of diseases ranging from 0 to 1, however, today the value has increased from 3 to 5 applications, increasing the costs of fungicides from 10 to 70 US$ respectively. This disease, year after year is responsible for at least 30 to 50% of loss in total production area, which in economic terms is between US$100 and US$150 million witch concerns to the use of agrochemicals and total yield loss per year (Condori, 2009).

In the 2007 winter cropping season the problems to control the rust emerged from a series of technical and climatic factors as (Condori, 2009):

- Soybean planted between harvests (April-May) generated the Asian rust inoculum that infected soybean fields planted in June until early winter season;
- The prolonged period of drought and the continuous moisture in the months of August and September stressed cultivation, focusing directly into the beginning of flowering that occurred at 65 to 70 days;
- The fungicides applied were exposed to critical climatic conditions as high temperatures (30-35°C) and low relative humidity (50-40%) which affected the residual effect and effectiveness of their control.

Those factors explain that this season (2007/08) was one of the most catastrophic, mainly in the north and east of the Santa Cruz de La Sierra due to continuous rains that prevented raising the winter planting crop, to perform the planting on summer and, the most important fact, the delay of fungicide applications, generating a "explosion" of the rust, forced farmers to make up 7 applications of fungicides per hectare. The economic losses quantified by the "Asociación de Productores de Oleaginosas y Trigo – ANAPO" (Association of Producers of Oilseeds and Wheat), exceed US$ 150 million, for the past two seasons (summer 2007/08 and winter 2008) (Condori, 2009).

The Bolivia are implementing the sanitary break in cities located in the Integrated Zone (Andrés Ibáñez Province, Warnes, Ichilo, Sara, Bishop Santiestevan and Guarayos), Expansion Area (Andres Ibanez, Chiquitos, Ñuflo Chavez, Guarayos) of the Santa Cruz state. This project will benefit more than 14,000 small, medium and large producers, of various nationalities and a planting area between the 700,000 to 1,000,000 hectares. Training, dissemination and sharing of technical and legal measures, through different media

available as workshops, seminars, television and radio messages is also referred in this project (Condori, 2009).
Fungicides recommended for control of Asian rust are bencimidazoles, triazols, triazol + triazol, triazol + benzimidazol and triazol + estrobilurina products. Those products were chose based on research results conducted by different agricultural companies and research institutions (Condori, 2009).

3.3 Brazil

Brazil is the second largest producer of soybeans. In the 2006/07 season, the culture occupied an area of 20.69 million hectares, which totaled a production of 58.4 million tons. The United States, the worldwide producer, accounted for the production of 86.77 million tons of soybean. The yield of soybeans in Brazil is 2,823kg per hectares, reaching about 3,000 kg/ha in Mato Grosso, the largest state producer (EMBRAPA, 2011).
The soybean is the crop which has the higher development in Brazil in the last three decades and accounts for 49% of grains area planted in the country. The grain is an essential component in the manufacture of animal feeds and the growing use food is increasing (MAPA, 2011). Data from the Ministry of Development show that soy has a major share of Brazilian exports. In 2006, were US$ 9.3 billion, representing 6.77% of total exported (EMBRAPA, 2011).
The Asian soybean rust was identified in Brazil in May of 2001 and spread quickly to the main producing regions, becoming a major problem for the national soybean producers. To propose solutions was created in September of 2004, the "Consórsio Antiferrugem - CAF" (Antirust Consortium). The consortium constituents are representative institutions of various soybean segments as foundations, universities, research institutes, representatives of entities of inputs manufacturers and farmer cooperatives. One of the aims of the Consortium is to bring the farmer all available information about the disease and enable him to handle it (Farias, 2009).
The CAF main information and communication vehicle is the consortium website: "www.consorcioantiferrugem.net" (in Portuguese) where the laboratories accredited update information about the disease outbreaks in all producing regions of Brazil during a season. In the system are recorded and presented a map of Brazil, the city of occurrence, date of detection, the fenological phases of culture and type of area (warning unit, commercial field, irrigated area etc). Thus, epidemics of soybean rust have been monitored and the spread of the disease are presented in real time at the consortium website, describing it as the main source of data for the record of events and the spread of the disease in Brazil (Spolti et al., 2009).
When the disease arrives, both farmers and technicians were not prepared to identify soybean rust. Factors such as dry climate, the symptoms likely with other diseases of end of cycle and because it was a new disease in the Americas, their identification was difficult and there was no species resistant to fungus attack (Constamilan, 2002 (2005)). It is estimated that over 60% of soya production in Brazil has been contaminated in the season of 2001/02, causing grains losses estimated at 569.2 thousand of tons or the equivalent of US$ 125.5 million (US$ 220.50/t) (Yorinori, 2004).
In the season 2002/03, the occurrence was different from the last season. In localities where the disease was severe in 2001/02, the high temperatures prevented, despite the high amount of rain, the development of the disease, except in Rio Grande do Sul and Santa Catarina, where the late cultivars were affected. But where the rust had not been reported earlier, favorable

climatic conditions and a new strain of *P. pachyrhizi* caused major losses. The states of Bahia, Goiás, Minas Gerais and Mato Grosso were severely affected (Yorinori, 2004).

However, despite the intensive campaigns to alert and guidance on methods for identification and control, held in 2002 and in January and February of 2003, through lectures, publications and other means of dissemination, the technical assistance and most producers were not prepared to control the rust. In many crops, the fungicide application was delayed due to lack of product and/or excessive rain which precluded the spraying (Yorinori, 2004). In this season the damage caused by the rust (amounting the grain losses, control expenses and revenues falling) were approximately US$ 1.29 billion (Soares et al. 2004).

The beginning of 2003/04 season was characterized by irregular rainfall and high temperatures, which probably not favored the outbreak of rust as expect. Moreover, the experience of loss in the previous crop left farmers in the areas previously affected readiness and "armed" for the chemical control. However in the southern region, the beginning of the harvest was characterized by mild temperatures and frequent rainfall, which favored the early emergence of *P. pachyrhizi*. The total damage caused by rust, in this year, adding the grain losses, control spending and falling revenue was approximately US$ 2.28 billion (Yorinori, 2004).

Among the crops of 2005/06 to 2008/09 were recorded, respectively, 1,369; 2,766; 2,107 and 2,880 reports of the occurrence of soybean rust. While there is an increase in the number of reports over the years, it is not possible to assert that the severity of epidemics is related to the number of outbreaks, since it is observed only presence of disease in crops, once in 2006/07, when they were registered the greatest losses in productivity caused by rust due to the higher disease severity, the number of reported outbreaks was lower than in 2008/09 when, according to regional information, the attack of the disease was not as severe as that year (Spolti et al., 2009).

Since the disease monitoring, the Asian rust was not observed before the month of October, whether in the commercial field or in units of alert. The progress of the number of reports of disease presents a sigmoid pattern with a logarithmic phase and a stationary phase when approaches the end of growing season. The maximum rate of increase, indifferent to the season, was observed between January and March (90 to 150 days after October 1st), this period can be defined as critical in the epidemics development, being responsible for the differentiation of the final number of focus reported in the cycle (Spolti et al., 2009).

At present, around 70 fungicides are registered in the Brazilian Ministry of Agriculture for managing soybean rust and many of these have been evaluated annually since 2003/2004 in a nationwide network of standardized coordinated by Embrapa Soja, a research unit of the Brazilian Agricultural Research Corporation (Godoy et al., 2010; Scherm et al., 2009). The fungicides registered for control of Asian soybean rust belong into two main groups: Triazoles and strobilirins (Godoy & Flausino, 2008).

3.4 Paraguay

The soybean in Paraguay is the main agricultural export item, with a market of 70% of national output in the form of grain. This is due to high charges imposed by the European Union, the main buyer, for other soy subproducts such as soybean oil. Today, Paraguay is the sixth soybean production in the world (preceded by USA, Brazil, Argentina, China and India) and has a weighted average of 2,600kg/ha, performance similar to Argentina and Brazil (CAS, 2008).

Since the appearance of Asian soybean rust in 2001 in Paraguay, there have been major changes in the soybean production system and it also contributed to better care for the crop, getting even better yields by protecting against various diseases of economic importance appellant in soybean. The productivity losses were very important in years when climatic conditions were favorable for the disease, especially during the breeding season and when constant rainfall recorded during the months of January and February, considered the most critical for the development of an epidemic Paraguay (Morel & Bogado, 2009).

In the first year where the disease was recorded yield losses were estimated, in the cultivars most affected, at more than 60%. On the next seasons, 2001/02 and 2002/03, the severity of the disease was not very important because of the drought but the late sown soybean crop showed severe losses of more than 50% of performance. This epidemic is especially observed when the rains season from the month of March to May (Morel & Bogado, 2009).

In all the years that the rust has been detected early, it was observed in plants of 30-35 days from sentinel plots in the region of Pirapó, considered an endemic area once is possible to detect the strong presence of the volunteer plant, being a fungus host, named Kudzu (*Pueraria lobata*), but the severity level ever has thrived in the vegetative phase. This demonstrates the importance of the survival of the disease during the winter, which has made a strong campaign of awareness among farmers aiming the elimination of inoculum source in areas with no winter crop, in order to avoid the primary infection in an early period of soybean cultivation (Morel & Bogado, 2009).

The crop of 2005/06, was the largest epidemic in the normal planting season, resulting in a loss of more than US$ 400 million. This strong impact due to multiple factors, neglect of producers, the time control and problems in application of the technology. The number of fungicide applications was a maximum of 5 and a minimum of 2. In the season of 2008/09, the rust incidence was reported again in a very early (second fortnight of October) in crops planted in September, but the severe drought that affected the whole area of soybean production allowed the progress of the disease (Morel & Bogado, 2009).

In rare cases and in regions where rains started to become evident, controls measures have been made, but around the country more than 1 application of fungicides should be done for each producer. This drought was so important that did not allow the progress of the disease throughout the production area, except in some regions of late-sown soybean (Morel & Bogado, 2009).

4. Disease control

Once commercial soybean cultivars used in the major soybean producing countries are susceptible to soybean rust, management of the disease is done using fungicides, although some cultural and crop management practices also may decrease disease risk at field and regional scales. Early research in Asia indicated that mancozeb and, to a more limited extent, the benzimidazole fungicides suppressed soybean rust but required three to five applications to be effective. The disease control was significantly improved after the introduction of the triazole fungicides (Scherm et al., 2009).

Scherm et al. (2009) studying the efficacy of several fungicides on a soybean crop in Brazil showed that triazole fungicides had significantly efficiency than strobilurins classes. The combination of triazoles with strobilurins improved disease control and yield gain compared with triazoles or strobilurins alone. However the combination of triazoles with a benzimidazole fungicide did not improve the desease control when compared with triazoles alone. They either conclude that the two fungicides with the best disease control efficacy were combinations of two active ingredients as flusilazole + carbendazim and azoxystrobin + cyproconazole.

The triazoles group acts to inhibit the ergosterol biosynthesis and have the primary site of action the C-14 demethylation and the strobilurins interfere with mitochondrial respiration by blocking electron transfer by the cytochrome bc1 complex, formulated alone or in mixtures (Godoy & Flausino, 2008).

Besides the fungicides application, other measures could be taken as the use of earliest varieties, seed at the beginning of recommended time for each region, avoid prolonging the period of sowing, inspect crops and verify if there are temperature and high humidity favorable to the pathogen (Reunião de Pesquisa da Soja da Região Sul, 2002).

One way to anticipate the presence of this fungus before it reaches the crop would be the establishment of sentinel plots in one or more locations, depending on the area of the property. These traps, seeded with 15 to 20 days in advance of the first crops are intensively monitored to identify the first symptoms. Once detected the presence of disease, the traps must be destroyed or heavily treated with an effective fungicide. From this initial detection, the commercial areas should be treated or monitored more carefully for making treatment decisions to be made (Yorinori & Lazzarotto, 2004).

Another method for disease control is the adoption of absence of living plants in the field of this culture denominated sanitary break. This technician aiming to reduce the amount of uredospores in the environment on off-season and then, inhibit the early attack to soybean plants, trough the smaller inoculum presence (Seixas & Godoi, 2007) and is adopted in many countries. The general rule is that all regions are forbidden to cultivate soybean in the period established and the remaining plants from the last crop should be eradicated with chemicals or other means. The producer who does not obey the sanitary break will be required to pay large fines. Another caution that the producer should be is to remove the soy plants that may grow due to grain that fell in the soil and germinate during the harvest.

Also, the kudzu (*Pueraria lobata*), a leguminous plant witch is highly susceptible to the Asian soybean rust, founded in Paraguay and Brazil, shown to be an efficient source of inoculum, presenting the first symptoms and fungal growth before the first crops of soybean (Yorinori & Lazzarotto, 2004). In those countries where kudzu is found, control policies of the rust also include the control of this plant.

5. References

Begenisic, F.; Ploper, L.D.; Ivancovich, A. (2004). *Roya de la soja: Características de la enfermedad*. Documento de Trabajo N° 1, Programa Nacional de la Roya de la Soja

Bromfield, K.R.; Melching, J.S.; Kingsolver, C. H. (1980). Virulence and aggressiveness of *Phakopsora pachyrhizi* isolates causing soybean rust. *Phytopathology*, Vol. 70, pp. 17-21

CAS - Consejo Agropecuario del Sur. (2008). *El Mercado de la soja en los países del Consejo Agropecuario del Sur*. Grupo II. Sistema de Información de Mercados y Pronósticos de Cosecha

Condori, M. (2009). Roya asiatica de la soya en Bolivia (Santa Cruz de La Sierra), In: *Reunião do Consórcio Antiferrugem Safra 2008-09*, Godoy, C.V.; Seixas, C.D.; Soares, R.M., pp. 47-54, Embrapa Soja, Londrina, Paraná, Brazil, Documentos 315

Constamilan, L.M.; Bertagnolli, P.F.; Yorinori, J.T. (2002). *Comunicado Técnico 96*, 10 fev. 2005, Available from: <www.embrapa.com.br>

Cui, D.; Zhang, Q.; Li, M.; Hartman, G. L.; Zhao, Y. (2010). Image processing methods for quantitatively detecting soybean rust from multispectral images. *Biosystems Engineering*, Vol. 107, pp. 186-193

Deverall, B. J.; Keogh, R. C.; McLeod, E. (1977). Responses of soybean to infection by, and to germination fluids from, urediniospores of *Phakopsora pachyrhizi*. *Transactions of the British Mycological Society*, Vol. 69, No. 3, pp. 411-415, 1977.

EMBRAPA – Empresa Brasileira de Pesquisa Agropecuária. *A soja*, 14 mar. 2011, Available from: <www.cnpso.embrapa.br/index.php?op_page=22&cod_pai=16>

Embrapa. 12 fev. 2005, Available from: <www.embrapa.com.br>

Embrapa. *Sistemas de produção*, 17 mai. 2004, Available from: <http://sistemasdeproducao.cnptia.embrapa.br/FontesHTML/Soja/SojaCentralBrasil2003/doenca.htm>.

Farias, J.R.B. (2009). Apresentação. In: *Reunião do Consórcio Antiferrugem Safra 2008-09*, Godoy, C.V.; Seixas, C.D.; Soares, R.M., Embrapa Soja, Londrina, Documentos 315

Godoy, C.V.; Canteri, M.G. (2004). Efeitos protetor, curativo e erradicante de fungicidas no controle da ferrugem da soja causada por *Phakopsora pachyrhizi*, em casa de vegetação. *Fitopatologia brasileira*, Vol. 29, pp. 97-101

Godoy, C.V.; Flausino, A.M. (2007). *Eficiência de fungicidas para o controle da ferrugem asiática da soja em Londrina e Tamarana, PR, na safra 2007/08*, Circular Técnica 57, Embrapa Soja, Londrina, Paraná, Brazil

Godoy, C.V.; Utiamada, C.M.; Silva, L.H.C.P. (2010). *Eficiência de fungicidas para o controle da ferrugem asiática da soja, Phakopsora pachyrhizi, na safra 2009/10*: resultados sumarizados dos ensaios cooperativos, Circular Técnica 80, Embrapa Soja, Londrina, Paraná, Brazil

Hartman, G.; Wang, T.; & Tschanz, A. (1991). Soybean rust development and the quantitative relationship between rust severity and soybean yield. *Plant Disease*, Vol. 75, No. 6, pp. 596-600

Huber, L.; & Gillespie, T.J. (1992). Modeling leaf wetness in relation to plant disease epidemiology. *Annual Review of Phytopathology*, Vol. 30, pp. 553-577

Kawuki, R.S.; Tukamuhabwa, P.; Adipala, E. (2004). Soybean rust severity, rate of rust development, and tolerance as influenced by maturity period and season. *Crop Protection*, Vol. 23, pp. 447-455

MAPA - Ministério da Agricultura, Pecuária e Abastecimento. *Soja*, 14 jan. 2011, Available from: <www.agricultura.gov.br >

Mendesa, R.K.; Carvalhal, R.F.; Stach-Machado, D.R.; Kubota, L.T. (2009). Surface plasmon resonance immunosensor for early diagnosis of Asian rust on soybean leaves. *Biosensors and Bioelectronics*, Vol. 24, pp. 2483-2487

Morel, W.; Bogado, N. (2009). Relato de la situación de la roya de soja en Paraguay. In: *Reunião do Consórcio Antiferrugem Safra 2008-09*, Godoy, C.V.; Seixas, C.D.; Soares, R.M., pp. 45-46, Embrapa Soja, Londrina, Paraná, Brazil, Documentos 315

Pivonia, S.; Yang, X. B. (2004). Assessment of the potential year-round establishment of soybean rust throughout the world. *Plant Disease*, Vol. 88, No. 5, pp. 523-525

Programa Nacional de la Roya de la Soja. *Acciones previstas para la campaña 2005/06*, 13 jan. 2011a, Available at: <www.minagri.gob.ar/SAGPyA/agricultura/roya_soja>

Programa Nacional de la Roya de la Soja. *Roya asiática de la soja: Escenario pasado, actual y perspectivas futuras en Argentina*, 13 jan. 2011b, Available from: <www.minagri.gob.ar/SAGPyA/agricultura/roya_soja>

Reunião de pesquisa de soja da região sul. (2002). *Indicações técnicas para a cultura da soja no Rio Grande do Sul e em Santa Catarina 2002/2003*, Fundacep/Fecotrigo, Cruz Alta, Rio Grande do Sul, Brazil

Scherm, H.; Christiano, R.S.C.; Esker, P.D.; Del Ponte, E.M.; Godoy, C.V. (2009). Quantitative review of fungicide efficacy trials for managing soybean rust in Brazil. *Crop Protection*, Vol. 28, pp. 774-782

Schmitz, H.F.; Grant, R. H. (2009). Precipitation and dew in a soybean canopy: Spatial variations in leaf wetness and implications for *Phakopsora pachyrhizi* infection. *Agricultural and Forest Meteorology*, Vol. 149, pp. 1621-1627

Seixas, C.D.S.; Godoy, C.V. (2007). Vazio sanitário: panorama nacional e medidas de monitoramento, In: Simpósio brasileiro de ferrugem asiática da soja. *Anais do simpósio brasileiro de ferrugem asiática da soja*, Embrapa Soja, Londrina, Paraná, Brazil

Soares, R.M.; Rubin, S.A.L.; Wielewicki, A.P.; Ozelame, J.G. (2004). Fungicidas no controle da ferrugem asiática (*Phakopsora pachyrhizi*) e produtividade da soja. *Ciência rural*, Vol. 34, No. 4, pp. 1245-1247

Spolti, P.; Godoy, C.V.; Del Ponte, E.M. (2009). Sumário da dispersão em larga escala das epidemias de ferrugem asiática da soja no Brasil em quatro safras (2005/06 a 2008/09). In: *Reunião do Consórcio Antiferrugem Safra 2008-09*, Godoy, C.V.; Seixas, C.D.; Soares, R.M., pp. 11-19, Embrapa Soja, Londrina, Paraná, Brazil, Documentos 315

Tremblay, A.; Hosseini, P.; Alkharouf, N.W.; Li, S.; Matthews, B.F. (2010). Transcriptome analysis of a compatible response by *Glycine max* to *Phakopsora pachyrhizi* infection. *Plant Science*, Vol. 179, pp. 183-193

Yorinori, J.T. (2002). Ferrugem asiática da soja (*Phakopsora pachyrhizi*): ocorrência no Brasil e estratégias de manejo. In: *II Encontro Brasileiro sobre Doenças da Cultura da Soja*, Resumos de palestras, pp. 47-54, Passo Fundo, Rio Grande do Sul, Brazil

Yorinori, J.T.; Lazzarotto, J.J. (2004). *Situação da ferrugem asiática da soja no Brasil e na América do Sul*, Documentos 236, Embrapa, Londrina, Paraná, Brazil

10

Soybean Cyst Nematode (*Heterodera glycines* Ichinohe)

Qing Yu
Eastern Cereal and Oilseed Research Centre
Agriculture and Agric-Food Canada, Ottawa, ON
Canada

1. Introduction

Soybean cyst nematode (SCN) (*Heterodera glycines* Ichinohe) is the most serious nematode pest on soybean in the world, infests most of the soybean producing countries of the world with the exception of west European countries and Oceania countries and causes up to 1.5 billions US$ economical loss according to some estimates (Wrather et al., 2001). The cyst nematode is also an international quarantined pest. Although it was first discovered and described in Japan in 1952, it is now widely believed originated in China as its host soybean was. In America, since the first report in North Carolina, USA in 1955, it has spread to 26 out of 28 soybean producing states, to the province of Ontario, Canada, and to the soybean producing countries of South America. Race of the nematode was recognized in 1954, and a total of 14 races were reported and widely distributed, especially in the USA, which has created a series of problems for developing resistant cultivars. As the climate change intensifies, it is likely that this nematode pest is going to spread to new soybean producing areas. Many resistant cultivars have been developed, especially in USA where resistant cultivars were developed using resistant parents selected from resistant plant introductions (PI) of the exotic accessions in the USDA soybean germplasm collection. These resistant PIs were collected from the oriental countries. Since the 80s, international seed companies like Pioneer, and Monsanto have been the driving force for the resistant cultivar development and marketing. SCN resistant cultivars alone used to be the solution for the control of the nematode pest for soybean production in USA. Because of the new emerging races and the shifting between existing races, resistant cultivars in many cases lose their usefulness dramatically. With the new realization that the agriculture biodiversity plays an essential role for pest management, new control methods have been developed and tested such as rotating with nonhost crops, planting multiline cultivars mixtures, using biological control agents, and applying green manure. There are a lot of literatures related with SCN, one book by Schmitt et al, 2004, and a review by Noel 1993 nevertheless are the excellent sources of information. A few non scientific aspects are also used for the synthesis of this paper. China, where the nematode is believed originated, becomes economically integrated into the world system at a pace never seen before. Soybeans there as a crop are shifting from vegetable to pulse, and to oil seeds. Soybean seeds are increasingly produced, controlled and marketed by a few international companies, as the result, fewer cultivars (more monocultures) are planted at a given year compared with the time when farmers used to get their seeds from all sorts of channels, and productions are in large scales. The climate change, especially the global warming caused by human activities is inevitably impacting the soybean production, and the soybean cyst nematode.

2. Taxonomic position

When the SCN was discovered, it was believed as a race of *Heterodera schachtii*, the sugar beet cyst nematode, since these 2 species are closely related biologically. Around that time, cyst nematodes were generally considered races of *H. schachtii*. In 1940 Franklin's comparative morphological studies led to many "races" being elevated to species. It was morphologically consistent that the morphological distinctions of the SCN led Ichinohe to elevate the race to a new species in 1952.

Phylum Nematoda,
　　　　Class Secernentea,
　　　　　　Order Tylenchida,
　　　　　　　　Suborder Hoplolaimina,
　　　　　　　　　　Superfamily Hoplolaimoidea,
　　　　　　　　　　　　Family Heteroderidae,
　　　　　　　　　　　　　　Genus *Heterodera*,
　　　　　　　　　　　　　　　　Species *Heterodera glycines* (Ichinohe 1952)

3. Morphology and identification

H. glycines is a typical cyst forming nematode within the family of Heteroderidae: characterized by sexual dimorphism: male is vermiform, while female lemon shaped (Figure 1). The brown cyst is the dead female with viable eggs inside. The second stage juvenile (J2) is vermiform, much smaller than the male. The length of the stylet, and the hyaline tail terminus of J2, and the characters of the vulva cone of the cyst are the most important characters for the identification (Table 1). More than one reference descriptions are usually required for comparison because there are variance among isolates from different crops and locations (Mulvey & Golden 1983, Wouts 1985, Golden 1986, Burrows & Stone 1985, Tylor 1975, Hesling 1978, Graney & Miller 1986).

Fig. 1. Morphology of *Heteridera glycines*: A: male; B: J2; C: head of J2; D: cyst; E: hyaline tail terminus

Character		Measurment (μ)	
Cyst		Average	Range
Fenestra	length	55	30-70
	width	42	25-60
Vulval slit	length	53	43-60
J2			
Body	length	440	375-540
Stylet	length	23	22-24
Tail	length	50	40-61
	hyaline length	27	20-30

Table 1. Measurement of *Heterodera glycines* (after Tylor, 1975, Graney & Miller, 1982)

4. Biology and life cycle

After the death of the female, the eggs are retained inside the hardened body (cyst), until suitable conditions arrive. The cysts can remain viable for several years in the soil. The eggs hatch to juveniles at stage 2, stimulated by exudates from the roots (Masamune et al., 1982). The 2nd stage juveniles (J2) of *H. glycines* are the only stage that the nematode can penetrate the root near the tip. The J2 once inside the root, become sedentary and establish a syncytic feeding site (Moore, 1984). J2 swells, and moults to J3, J4, and become adults (Wyss and Zunke, 1992). The life cycle is usually from 21 to 24 days (Fig. 2). Time required for the nematode to complete its life cycle is usually from 20-25 days at 20-24 °C, the lower the temperature is, the longer the time it takes to finish its life cycle (Melton et al., 1986).

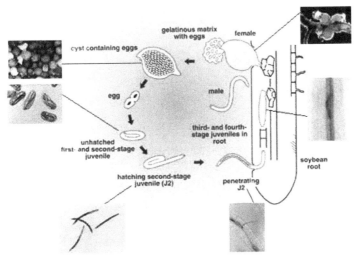

Fig. 2. Life cycle of *Heterodera glycines* (courtesy of Dirk Charlson, Iowa State University)

5. Distribution

It is impossible to know exactly the source(s) of infestation of the nematode species. More and more people believe that *H. glycines* is very likely a native of China, for two compelling evidences: 1. One of the most important hosts for the nematode the soybean originated (domesticated 5,000 years ago, the early stage of the Chinese civilization) (Qiu et al., 2011, Liu et al., 1997) in China, and 2. Most of the resistant cultivars used today have their roots in cultivars from China (Bernard et al., 1988). Probably the spreading and the pathway of the nematode followed footsteps of its host soybean (Fig. 3). The soybean was introduced to Japan, Korea around 300 AD, the nematode was discovered in 1915 (Hori), and later was described by Ichinohe in 1952 with the type locality in Hokkaido. It was first found in the United States in 1954 (Winstead et al., 1955) and spread with the expansion of soybean growing areas such as in Canada (Anderson et al., 1988). The nematode was also found in Colombia in the 1980s, and more recently in the major soybean producing areas in Argentina and Brazil (Mendes & Dickson 1992). SCN has also been reported from Iran and Italy (Fig. 3).

Africa: Egypt (unconfirmed)

Asia: China (Anhui, Hebei, Hubei, Heilongjiang, Henan, Inner Mongolia, Jiangsu, Jilin, Liaoning, Shanxi, Shandong), Indonesia (Java), Korean peninsula, Japan, Taiwan (unconfirmed), Russia (Amur District in the Far East).

North America: Canada (Ontario), USA (Alabama, Arkansas, Delaware, Florida, Georgia, Illinois, Indiana, Iowa, Kansas, Kentucky, Louisiana, Maryland, Minnesota, Michigan, Mississippi, Missouri, Nebraska, New Jersey, North Carolina, North Dakota, Ohio, Oklahoma, Pennsylvania, South Carolina, South Dakota, Tennessee, Texas, Virginia and Wisconsin).

South America: Argentina, Brazil, Chile, Columbia, Ecuador.

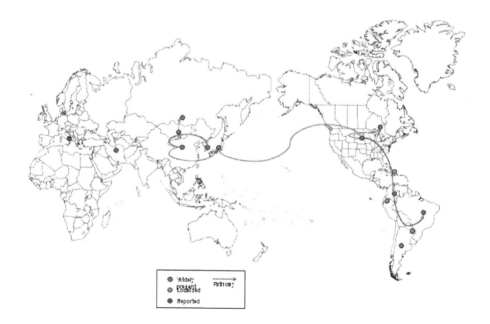

Fig. 3. Distrubition map of *Heterodera glycines* in the World.

6. Disease symptoms

"Yellow dwarf" is a good description of the above ground symptoms when soybeans are infested with the soybean cyst nematode. When soybeans are heavily infested, the plants usually become stunt (Fig. 4). Low level infestation usually does not produce obvious symptoms above ground. Belowground symptoms include poorly developed and darkened roots, reduced nodule formation.

Fig. 4. Above ground symptoms of soybeans infested by soybean cyst nematode

7. Spread

7.1 Canada

Since it was first identified in 1987, SCN has been identified in 12 counties in Ontario, Canada. Infected counties include Essex, Kent, Lambton, Elgin, Perth, Haldimand-Norfolk, Middlesex, Glengarry, Prescott, Stormont, Huron and Oxford. It is obvious that *H. glycines* has been spreading north and northeast wards (Fig. 5). It would be difficult to exclude the possibility that climate change is the cause, or at least one of the causes. This finding proves what Boland et al. 2004 has predicated. It is likely to spread along the St. Lawrence seaway towards the Maritime provinces as new cultivars suitable for the cold climate being developed and planted in the region.

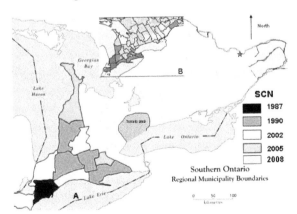

Fig. 5. Distribution of *Heterodera glycines* in Ontario, Canada from 1987 to 2007 (insert B: soybean growing area in Ontario).

7.2 USA

Since *H. glycines* was first discovered in North Caroline in 1955. Within the next 6 years, it had been reported in 7 states along the Mississippi river (Riggs, 2004). Today, it has spread to 29 soybean producing states in USA, as far northeast as the state of New Jersey (Riggs, 2004) (Fig. 6).

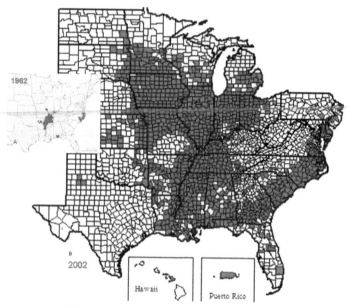

Fig. 6. Distribution map of *Heterodera glycines* in USA: A: distribution in 1962; B: distribution in 2002.

8. Host

The soybeans are the most economic important host for *H. glycines*. It has a broad host range, up to 100 plant species world wide, some selected common hosts are listed in Table 2 (Baldwin & Mundo-Ocampo, 1991), especially some legumes including beans, vetch, clover, and pea. It also attacks many species of weeds (Riggs & Hamblen, 1966). These weeds in the fields must be taken into consideration for any management measures. Some nonhosts are also listed that may be used in rotation.

9. Race

Variability in virulence among populations of soybean cyst nematode was recognized by researchers soon after the nematode was discovered in USA. The term race has been adopted by an *ad hoc* committee of the Society of Nematologists to designate the SCN populations differ in their ability to develop on a set of differential soybean cultivars (Table 3). This scheme was developed in 1970 (Golden et al., 1970) and later was refined by (Riggs & Schmitt 1988). The scheme defines 16 possible races. In 2002, HG type was introduced to more adequately define the diversity in virulence phenotypes (Niblack et al., 2002).

Host species	Non-host
Soybeans	Alfalfa
Green beans	Barley
Snap beans	Canola
Dry beans	Red clover
Red beans	White clover
Lima beans	Ladino clover
Mung beans	Oats
Bush beans	Rye
Adzuki beans	Sorghum
Graden peas	Wheat
Cowpeas	
Corn	
Vetch (common, hairy)	
Lespedeza (common, Korean, round bush, sericea)	
Birdsfoot-trefoil	
Sweetclover	
White lupines	

Table 2. A list of hosts and nonhosts for *Heterodera glycines*.

RACE	Pickett	Peking	PI 88788	PI 90763
1	-	-	+	-
2	+	+	+	-
3	-	-	-	-
4	+	+	+	+
5	+	-	+	-
6	+	-	-	-
7	-	-	+	+
8	-	-	-	+
9	+	+	-	-
10	+	-	-	+
11	-	+	+	-
12	-	+	-	+
13	-	+	-	-
14	+	+	-	+
15	+	-	+	+
16	-	+	+	+

A "+" rating is given if the number of females produced by a soybean cyst nematode population on the soybean cultivar is equal to or greater than 10% of the number produced on the susceptible cultivar Lee. If it is less than 10%, a "-" rating is given.

Table 3. Race classification scheme for *Heterodera glycines*.

Races 1, 2, 3, 4 were described in 1970 (Golden et al., 1970), race 5 in 1979 from Japan (Inagaki, 1979), and race 7 from China (Chen et al., 1988). Today a total 14 races have been reported (Table 4). Only races 12 and 16 have not been reported. USA has the most races. Race 3 is the most common race in the world.

With relative short history of growing soybean (100-200 years), the fact that USA has the highest number of races for the nematode indicates that the races were probably the results of its widespread planting of the resistant cultivars. Argentina (Doucet et al., 2008) and Brazil (Dias et al., 1998) both have a very short history of growing soybean (less than 50 years) have 6, and 9 races respectively. Very likely these races have been imported through seed stocks and on used machinery from several sources in USA. In comparison, the oriental countries, Korea (Kim et al., 1998), Japan (Ichinohe, 1988) and China (Liu et al., 1998) where soybean has been growing for several thousands years, have relative fewer races, because the modern practice of resistant cultivar selection and development is relatively new. Races 8, 11, 13, and 15 have only been found in USA. Lack of race 4 in both Japan and Korean was a surprise.

Country	Races	Total Races	Dominate Race
Argentina	1,3,5,6,9,14	6	3
Brazil	1,2,3,4,5,6,9,10,14	9	3
Canada	1,2,3,5,6	5	3
China	1,2,3,4,5,6,7,14	8	3 Northern, 4 Southern
Japan	1,3,5	3	3
Korea	1,3,5,6	4	3
USA	1,2,3,4,5,6,7,8,9,10,11,13,14,15	14	3

Table 4. Distribution of races of *Heterodera glycines* in the world

In USA, the most prevalent race for the northern states is the race 3, while the race 6 is the most common race. That difference of variability was noted along the 35 °N latitude line. Niblack and Riggs (2004) postulated that is likely caused by the history of using different resistant cultivars.

In central China lies between the Yellow and Yangtze rivers which include the province of Shandong, Anhui, Jiangsu, and Henan, and Shanxi, where the soybean may have originated, where the soybeans are planted in the summer and harvested in the fall (short growing season), the most common race is race 4 (Lu et al., 2006), race 3 has not been reported in that region. In northern China includes the province of Heilongjiang, Jilin, Liaoning, and Inner Mongolia where the soybeans are sown in spring and harvested in fall (long growing season) the prevalent race is race 3 and race 4 has only been reported in one case (Dong et al. 2008) (Fig. 7). China is only country that has these 2 races geologically separated. In the

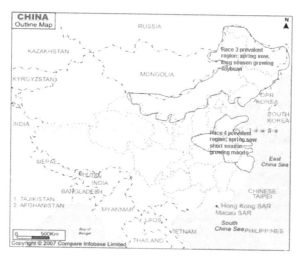

Fig. 7. The race 3 prevalent region and the race 4 prevalent region of *Heterodera glycines* in China

race classification scheme, the race 3 and the race 4 are opposite to each other, all the 4 cultivars are resistant to the race 3 but susceptible to the race 4. The cultivars have their resistant genes rooted in the cultivars collected from the northern region of China where the race 3 is the prevalent. It is likely that these 2 races are the 2 original races and are native to these 2 regions respectively in China, with race 4 is the ancestral to the race 3, since the cradle of the ancient Chinese civilization happened to be in the race 4 region, and ancient Chinese civilization was an agricultural civilization. The north region (race 3 region) was not an ancient agricultural area, rather a nomadic region.

10. Management

10.1 Resistant cultivar

A search for sources of SCN resistance led to the evaluation of large number of the plant introduction (PI) from among exotic accession in the USDA soybean germplasm collection. Five accessions were selected as parents. The first SCN resistant cultivar "Picket" with Peking as the source for resistance, was breed and released in 1966 in USA (Brim & Ross, 1966), hundreds have been developed for all the Maturity Groups. In 1970, field populations were found readily reproduce on Picket, and Peking. That finding led to the search for another source of resistance. Chosen from USDA soybean germplasm collection, PI 88788 was used for the development of "Bedford" (Hartwig & Epps, 1978). With the widespread deployment of the SCN resistant cultivars, new races emerged. This has been the story for SCN resistant cultivar development in USA. Since the 80s, the soybean seed breeding has been transferred from public institutes to private company. At much later stage, some accessions of the Chinese soybean germplasm collection were identified have resistance. At present, Roundup ready cultivars developed by Monsanto are the most widely used cultivars in USA, and elsewhere.

Relying on few resistant cultivars alone for SCN control had been proven misguided, as the high number and shifting of the races in USA indicated (Young, 1992).

In North America, the basic management tactics of planting resistant cultivars at different fashions, and rotating with non-hosts will continue to be the main methods to manage the SCN problem, even though the tactics face great challenges of the shitting of the nematode races, and of the uneconomical of the non-hosts (Niblack & Chen, 2004).

10.2 Using mutiline cultivars for SCN management

Probably, it is hard to argue against the fact that monoculture farming has been one of causes for disease and pest epidemics, the best example is without doubt the Irish Potato Famine caused by the potato late blight disease (*Phytophthora infestans*). Recent cases of other invasive alien species such as the Dutch Elm disease also remind us that biodiversity is very important in fighting pests and diseases. The crop biodiversity used to be a norm practice before the modern agriculture (few cultivars, and a few pesticides), each farmer had to grow different kinds of crops for all household needs (grains, vegetable, and others). The usefulness of mixture of multiline cultivars and cultivars mixtures for disease control has been well documented (Mundt, 2002, Wolfe, 1985). The recent successful cases of using multiline cultivars or cultivar mixtures for controlling diseases, such as potato late blight on potato (Garrett & Mundt, 1999), on barley (Wolfe et al. 1981), on rice in China (Zhu et al., 2000) demonstrated that the practical difficulties associated with the mixtures have been overestimated. This concept has not been carefully tested for SCN control. In the few tested cases, the mixtures were not superior to the resistant cultivars in terms of their yield increasing

(Young & Hartwig, 1988). More studies are recommended. As Mundt (2002) demonstrated that for biodiversity to be functional, there must be an appropriate match between the resistant genes in a mixture and the virulence genes present in the target pathogens or parasites.

10.3 Cover crop

Cover crops are commonly used to prevent soil erosion. These crops are usually planted in rotation with primary crops. When the cover crops are incorporated into the soil at the certain stage of the growing season, this practice is being referred as green manure. A major benefit obtained from green manures is the addition of organic matter to the soil, which increases the food supply for macro, and micro organisms in the soil resulting increased biodiversity in soil. There is a lot of information on the benefit effects of soil biodiversity on disease control (Brussaard et al., 2007).

This agriculture practice with certain crops which contain nematicidal compounds is especially interesting. Marigold, especially French marigold (*Tagetes patula*) has been shown reduced the populations in soil of several root-knot nematodes, and root lesion nematodes (Motsinger et al., Ploeg, 2000, Pudasaini, 2007). Castor beans, sesame, Sudan grass, sorghum, and Crucifers have all shown are toxic against plant parasitic nematodes. Among them, plants from Brassica have received considerable attention for their possibility in controlling plant parasitic nematodes by incorporating them into soil (Mojtahedi et al., 1993, Potter et al., 1998). The principle reason is that glucosinolates which exist in these plants convert upon decomposition to isothiocyanates, a group of chemicals proven to have a wide spectrum of biological activities, including nematicidal activity (Brown & Morra 1997), a few these chemicals are volatile, the practice has been referred "Biofumigation". Among these converted isothiocynates, allyl isothiocyanate (AITC) has been proven as being the most toxic against *H. glycines* (Lazzeri et al. 1993). AITC is the decomposition product of allyl glucosinolate (generally called sinigrin), which exists in plants of *Armoracia lapathifolia, Brassica carinata, B. juncea, B. napus, B. oleracea,* and *Peltaria alliacea* (Brown & Morra, 1997). Among them, mustards have been cited most promising, especially the oriental mustard (Brassica juncea) which contains highest concentration of Ally isothiocynate (AITC) in plant (Tsao et al., 2000). AITC toxicity was found highly selective, was highly toxic against J2 of *H. glycines*, but less toxic on

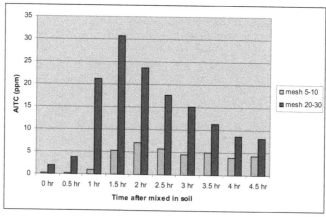

Fig. 8. Effect of particle size of mustard materials on AITC releasing in soil

free-living nematodes, AITC also inhibited the egg hatching of the nematode (Yu et al., 2005). Some materials from this oriental cultivar have been demonstrated effective in reducing population of *Pratylecnhus penetrans* in soil (Yu et al., 2007 a, 2007b).

Recently using mustard such as oil radish, or other mustard related crops as a cover crop for controlling *H. glycines* have tested, but the results have not been conclusive. The potential factors that caused the inconsistency includes: 1) targeted nematode species; 2) mustard varieties; and 3) environmental factors. In another study we found that the particle size had dramatic effect on releasing the AITC in to the soil (Fig. 8). It is likely that with a mustard variety of high AITC concentration, and plant tissue macerated to very fine particles, mustard crops as cover crops for the SCN control can be an effective method.

11. Concluding remarks

The soybean cyst nematode is more likely going to spread to new soybean growing areas around the world as the climate change intensifies, and as the world becomes more integrated. The soybean seeds are more than ever developed and marketed by a few international companies, the soybean farming practices in the world will become more and more uniform, less diverse unfortunatly. New races could emerge. This creates greater challenges for managing the SCN.

The whole genome sequencing project of *H. glycines* has been completed by Monsanto Company and Divergence. The sequencing information although has been submitted to Genbank, it remains inaccessible to the public. DOE Join Genomic Institute led by Kris Lambert and Matthew E. Hudson (Univ. of Illinois at Urbana-Champaign) is in the process of sequencing the pest as well, in a hope that it will lead us to learn more about the races, and to find new ways for the controlling of the pest.

There are a few soybean germplasm collections in the world with the USDA collection being the largest, the holding information is accessible to the public. The Chinese soybean germplasm collection holds 6644 accessions of *Glycine soja*, a potential rich pool of source of resistance. Collaborations between the collections such as sharing information and germ lines are essential. The management of SCN must not rely on a few resistant cultivars. An integrated approach involving several cultivars, rotating with nonhosts, and cultivation practices that encourage biodiversity in the soil must be the future.

12. Acknowledgment

Appreciation goes to Dr. W. Ye of Nematode Assay Section, Agronomic Division, North Carolina Department of Agriculture & Consumer Services, NC, USA for providing some of the pictures.

13. References

Anderson, T. R., Welacky, T.W., Olechowski, T. H., Ablett, G., & Ebsary, B.A. (1988). First report of *Heterodera glycines* in Ontario, Canada. *Plant Disease* 72: 453.

Baldwin, J. G. & Mundo-Ocampo, M. (1991). Heteroderinae, cyst-and non-cyst- forming nematodes, In: *Manual of Agricultural Nematology*, WR Nickle, (Ed.), pp. 275-362, New York: Marcel Dekker Inc.

Boland, G.J., Melzer, M.S., Hopkin, A., Higgins, V., & Nassuth A. (2004). Climate change and plant disease in Ontario. *Canadian Journal of Plant Pathology* 26: 335-350

Brim, C.A., & Ross, J.P. (1966). Registration of Pickett soybeans. *Crop Science* 6: 305.

Brown P. D. & Morra, M.J. (1997). Control of soil-borne plant pests using glucosinolate-containing plants. *Advances in agronomy* 61:167-231.

Brussaard, L., de Ruiter, P., & Brown, G.G. (2007). Soil biodiversity for agricultural sustainability. *Agriculture, Ecosystems and Environment* 121:233-244.

Burrows, P.R. & Stone, A.R. (1985). *Heterodera glycines. CIH Descriptions of Plant- Parasitic Nematodes* No. 118. CAB International, Wallingford, UK.

CABI, http://www.cabi.org/isc/?compid=5&dsid=27027&loadmodule=datasheet&page =481&site=144.

Chen, P.H., Zhang, D.S. & Chen, S.Y. (1988). First report on a new physiological race (race 7) of soybean cyst nematode (*Heterodera glycines*). *Journal of the Chinese Academy of Agricultural Sciences* 20: 94.

Dias, W.P., Silva, J.F.V, & Pereira, J.E. (1998). Survey of *Heterodera glycines* races in Brazil, harvest 1997/98. *Nematologia Brasileira* 22: 33.

Dong, L.M., Xu, Y.L., Li, C.J., Pan, F. J., Xie, Y.J., Han, Y.B., Teng, W.L., & Li, W.B. (2008). Cyst density and subspecies identification of soybean cyst nematode in Heilongjiang province. *Chinese Journal of Oil Crop Sciences* 30:108-111.

Doucet, M. E., Paola L. & Coronel, N. (2008). The soybean cyst nematode *Heterodera glycines* Ichinohe in Argentina, In: *Integrated Management and Biocontrol of Vegetable and Grain Crops Nematodes*, A. Ciancio and K. G. Mukerji (ed.),. Springer the Netherland.

Garrett, K.A., & Mundt, C. C. (1999). Epidemiology in mixed host populations. *Phytopathology* 89: 984-990.

Golden, A. M. (1986). Morphology and identification of cyst nematodes, In: *Cyst Nematodes* Lamberti F, Taylor CE, (Ed.), pp. 23–45, Plenum press, New York, USA.

Golden A.M., Epps J.M., Riggs R.D., Duclos L.A., Fox J.A., & Bernard R.L. (1970). Terminology and identity of infraspecific forms of the soybean cyst nematode (*Heterodera glycines*). *Plant Disease Reporter* 54: 544-546.

Graney, L.S.O. & Miller, L.I. (1982) Comparative morphological studies of *Heterodera schachtii* and *H. glycines*. In: *Nematology in the Southern region of the United States* Riggs, R.D. (Ed.) pp 96-107, Southern Cooperative Series Bulletin No. 276

Inagaki, H. (1979). Race statues of five Japanese populations of *Heterodera glycines. Japanese Journal of Nematology* 9: 1-4.

Hartwig, E.E. & Epps, J.M. (1978). Registration of Bedford soybeans. *Crop Science* 18:915.

Hesling, J.J. (1978). Cyst nematodes: morphology and identification of *Heterodera, Globodera* and. *Punctodera*. In: *Plant nematology*, Southey, J.F. (Ed.), pp 125-155, HMSO, London, UK.

Hori, S. (1915). Phytopathological notes: sick soil of soybean caused by a nematode (in Japaneses). *Byotyugai-Zasshi* 2: 927-930.

Ichinohe, M. (1952). On the soy bean nematode, *Heterodera glycines* n. sp., from Japan. *Oyo-Dobutsugaku-Zasshi* 17: 4.

Ichinohe M. (1988). Current research on the major nematode problems in Japan. *Journal of Nematology* 20: 184-190.

Kim, D.G., Lee, J.K., & Lee, Y.K. 1999. Distribution of races of soybean cyst nematode in Korea. *Korea Journal of Applied Entomology* 38: 249-253.

Liu, X.Z., Li, J.Q., & Zhang, D.S. (1997). History and statues of soybean cyst nematode in China. *International Journal of Nematology* 7: 18-25.

Lu, W.G., Gai, J.Y., & Li, W.D. (2006). Sampling survey and identification of races of soybean cyst nematode (*Heterodera glycines*) in Huang-Huai valley. *Scientia Agricultura Sinica* 39: 306-312.

Masamune, T., Anetai, M., Takasugi, M. & Katsui, N. (1982). Isolation of a natural hatching stimulus, glycinoeclepin A, for the soybean cyst nematode. *Nature* 297:495-496.

Melton, T. A. , Jacobsen, B. J. , & Noel, G. R. (1986). Effects of Temperature on Development of *Heterodera glycines* on *Glycine max* and *Phaseolus vulgaris*. *Journal of Nematology* 18: 468-474.

Mendes, M.L., & Dickson, D.W. (1992). *Heterodera glycines* found on soybean in Brazil. *Journal of Nematology* 24: 606.

Mojtahedi, H., Santo, G. S., Wilson, J. H. & Hang, A. N. (1993). Managing *Meloidogyne chitwoodi* on potato with rapeseed as green manure. *Plant Disease* 77:42-46.

Morre, W. F. (1984). *Soybean cyst nematode.* Mississippi Cooperative Extension Service.

Motsinger, R. E., Moody, E. H. & Gay, C. M. (1977). Reaction of certain French marigold (*Tagetes patula*) cultivars to three *Meloidogyne* spp. *Journal of Nematology* 9: 278.

Mulvey, R.H. & Golden, A.M. (1983). An illustrated key to the cyst-forming genera and species of Heteroderidae in the Western Hemisphere with species morphometrics and distribution. *Journal of Nematology* 15: 1-59.

Mundt, C.C. (2002). Use of multiline cultivars and cultivar mixtures for disease management. *Annual Review of Phytopathology* 40: 381-410.

Niblack, T.L., Arelli, P. R., Noel, G.R., Opperman, C.H., Orf, J., Schmitt, D.P., Shannon, J.G., & Tylka, G.L. (2002). A revised classification scheme for genetically diverse populations of *Heterodera glycines*. *Journal of Nematology* 34:279-288.

Niblack, T. L., & Chen S. Y. (2004). Cropping systems and crop management practices, In: *Biology and Management of the Soybean Cyst Nematode* pp. 181–206, D. P. Schmitt, J. A. Wrather, & R. D. Riggs, (Eds.), Schmitt & Associates of Marceline. Marceline, Missouri, USA.

Noel, G. (1993). *Heterodera glycines* in Soybean. *Nematologia Brasileira* 17: 103-121.

Ploeg, A. T. (2000). Effects of amending soil with *Tagetes patula* cv. Single Gold on *Meloidogyne incognita* infestation on tomato. *Nematology* 2: 489-493.

Potter, M. Davies, J., K. & Rathjen, A. J. (1998). Suppressive impact of glucosinolates in Brassica vegetable tissues on root lesion nematode *Pratylenchus neglectus*. *Journal of Chemical Ecology* 24:67-80.

Pudasaini, M.P., Viaene, N., & Moens, M. 2006. Effect of marigold (Tagetes patula) on population dynamics of *Pratylenchus penetrans* in a field, Nematology, Vol. 8 (4): 477-484.

Qiu, L.J., Chen, P.Y., Liu, Z. X., Li, Y. H., Guan, R. X., Wang, L. H., & Chang, R. Z. (2011). The worldwide utilization of the Chinese soybean germplasm collection. *Plant Genetic Resources: Characterization and Utilization* 9: 109-122.

Riggs, R.D. (2004). History and distribution. In: *Biology and Management of the Soybean Cyst Nematode* pp. 9-40, D. P. Schmitt, J. A. Wrather, & R. D. Riggs, (Eds.), Schmitt & Associates of Marceline. Marceline, Missouri, USA.

Riggs, R. D. , & Hamblen, M. L. (1966). Further studies on the host range of the soybean cyst nematode. *Arkansas Agriculture Experiment Station Bulletin* 718:1-19.

Riggs, R.D. & Schmitt, D.P. (1988). Complete characterization of the race scheme for *Heterodera glycines*. *Journal of Nematology* 20: 392-395.

Schmitt, D. P., Wrather, J. A. & Riggs, R.D. (2004). Biology and Management of the soybean Cyst Nematode (Second Edition). Schmitt & Associates of Marceline, Marceline, Missouri, USA

Shannon, J.G., Arelli, P.R., & Young, L.D. (2004). Breeding resistance and tolerance. In: *Biology and Management of the Soybean Cyst Nematode* pp. 155-180, D. P. Schmitt, J. A. Wrather, & R. D. Riggs, (Eds.), Schmitt & Associates of Marceline. Marceline, Missouri, USA.

Taylor, A.L. (1975). Identification of soybean cyst nematodes for regulatory purposes. Proceedings. *Soil and Crop Science Society of Florida* 34: 200-206

Tsao, R., Yu, Q., Friesen, I., Potter, J., & Chiba, M. (2000). Factors affecting the dissolution and degradation of oriental mustard-derived sinigrin and allyl isothiocyanate in aqueous media. *Journal of Agriculture and Food Chemistry* 48: 1998-1902.

Young, L.D. (1992). Croping sequencing effects on soybean and *Heterodera glycines*. *Plant Disease* 76:78-81.

Young, L.D., & Hartwig, E.E. (1988). Selection pressure on soybean cyst nematode from soybean cropping sequences. *Crop Science* 28:845-847.

Winstead, N.N., Skotland, C.B. & Sasser, J.N. (1955). Soybean cyst nematode in North Carolina. *Plant Disease Reporter* 39: 9-11.

Wolfe, M.S. (1985). The current status and prospects of multiline cultivars and variety mixtures for disease resistance. *Annual Review of Phytopathology* 23: 251-271.

Wolfe, M.S., Barrett, J.A., & Jenkin, J.E.E. (1981). The use of cultivar mixtures for disease control. In: *Strategies for the control of cereal diseases*, J.F., Jenkyn, R.T., Plumb (Eds.), pp. 73-80 Blackwell Scientific Publications, Oxford, UK.

Wouts, W. M. (1985). Phylogenetic classification of the family Heteroderidae (Nematoda: Tylenchida). *Systematic parasitology* 7:295-328.

Wrather, J. A., Anderson, T. R., Arsyad, D. M., Tan, Y., Ploper, L. D., Porta-Puglia, A., Ram, H. H., & Yorinori, J. T. (2001). Soybean disease loss estimates for the top ten soybean-producing countries in 1998. *Canadian Journal of Plant Pathology* 23:115-121.

Wyss, U., & U. Zunke. (1992). Observations on the feeding behaviour of *Heterodera schachtii* throughout development, including events during moulting. *Fundamental and Applied Nematology* 15:75-89.

Yu, Q., Tsao, R., Chiba, M. & Potter, J. (2005). Selective nematicidal activity of allyl isothiocyanate. *Journal of Food, Agriculture and Environment* 3: 218-221.

Yu, Q., Tsao, R., Chiba, M., & Potter, J. (2007a). Elucidation of the nematicidal activity of bran and seed meal of Oriental mustard (*Brassica juncea* L.) under controlled conditions. *Journal of Food, Agriculture and Environment* 5 (3&4): 374-379.

Yu, Q., Tsao, R., Chiba, M., & Potter, J.W. (2007b). Oriental mustard bran reduces Pratylenchus penetrans on sweet corn. *Canadian Journal of Plant Pathology* 29:421-426.

Zhu, Y. Y., Chen, H. R., Fan, J. H., Wang, Y. Y., Li, Y., Chen, J. B., Fan, J. X., Yang, S. S., Hu, L. P., Leung, H., Mew, .T W., Teng, P. S., Wang, Z. H., & Mundt C C. (2000). Genetic diversity and disease control in rice. Nature 406: 718-722.

Permissions

The contributors of this book come from diverse backgrounds, making this book a truly international effort. This book will bring forth new frontiers with its revolutionizing research information and detailed analysis of the nascent developments around the world.

We would like to thank Hany A. El-Shemy, Ph.D., for lending his expertise to make the book truly unique. He has played a crucial role in the development of this book. Without his invaluable contribution this book wouldn't have been possible. He has made vital efforts to compile up to date information on the varied aspects of this subject to make this book a valuable addition to the collection of many professionals and students.

This book was conceptualized with the vision of imparting up-to-date information and advanced data in this field. To ensure the same, a matchless editorial board was set up. Every individual on the board went through rigorous rounds of assessment to prove their worth. After which they invested a large part of their time researching and compiling the most relevant data for our readers. Conferences and sessions were held from time to time between the editorial board and the contributing authors to present the data in the most comprehensible form. The editorial team has worked tirelessly to provide valuable and valid information to help people across the globe.

Every chapter published in this book has been scrutinized by our experts. Their significance has been extensively debated. The topics covered herein carry significant findings which will fuel the growth of the discipline. They may even be implemented as practical applications or may be referred to as a beginning point for another development. Chapters in this book were first published by InTech; hereby published with permission under the Creative Commons Attribution License or equivalent.

The editorial board has been involved in producing this book since its inception. They have spent rigorous hours researching and exploring the diverse topics which have resulted in the successful publishing of this book. They have passed on their knowledge of decades through this book. To expedite this challenging task, the publisher supported the team at every step. A small team of assistant editors was also appointed to further simplify the editing procedure and attain best results for the readers.

Our editorial team has been hand-picked from every corner of the world. Their multi-ethnicity adds dynamic inputs to the discussions which result in innovative outcomes. These outcomes are then further discussed with the researchers and contributors who give their valuable feedback and opinion regarding the same. The feedback is then collaborated with the researches and they are edited in a comprehensive manner to aid the understanding of the subject.

Apart from the editorial board, the designing team has also invested a significant amount of their time in understanding the subject and creating the most relevant covers. They scrutinized every image to scout for the most suitable representation of the subject and create an appropriate cover for the book.

The publishing team has been involved in this book since its early stages. They were actively engaged in every process, be it collecting the data, connecting with the contributors or procuring relevant information. The team has been an ardent support to the editorial, designing and production team. Their endless efforts to recruit the best for this project, has resulted in the accomplishment of this book. They are a veteran in the field of academics and their pool of knowledge is as vast as their experience in printing. Their expertise and guidance has proved useful at every step. Their uncompromising quality standards have made this book an exceptional effort. Their encouragement from time to time has been an inspiration for everyone.

The publisher and the editorial board hope that this book will prove to be a valuable piece of knowledge for researchers, students, practitioners and scholars across the globe.

List of Contributors

Cristiane Fortes Gris
Federal Institute of Southern Mines, Brazil

Edila Vilela de Resende Von Pinho
Federal University of Lavras, Brazil

Guajardo Jesús, Morales Elpidio, López Francisco, Quintero Cristina, Compean Martha, Noriega María-Eugenia, González Jesús and Ruiz Facundo
Universidad Autónoma de San Luis Potosí, Mexico

Kuniyuki Saitoh
Okayama University, Japan

Takuji Ohyama
Faculty of Agriculture, Niigata University, Japan
Quantum Beam Science Directorate, Japan Atomic Energy Agency, Japan

Hiroyuki Fujikake, Hiroyuki Yashima, Shinji Ishikawa, Kuni Sueyoshi and Norikuni Ohtake
Faculty of Agriculture, Niigata University, Japan

Satomi Ishii and Shu Fujimaki
Quantum Beam Science Directorate, Japan Atomic Energy Agency, Japan

Sayuri Tanabata
Agricultural Research Institute, Ibaraki Agricultural Center, Japan

Takashi Sato
Faculty of Bioresource Sciences, Akita Prefectural University, Japan

Toshikazu Nishiwaki
Food Research Center, Niigata Agricultural Research Institute, Japan

S. Calvo and M. L. Salvador
National University of Córdoba

S. Giancola, G. Iturrioz, M. Covacevich and D. Iglesias
National Institute of Agricultural Technology –INTA, Argentina

Vlado Kovacevic and Manda Antunovic
University J. J. Strossmayer in Osijek, Faculty of Agriculture, Croatia

Aleksandra Sudaric
Agricultural Institute Osijek, Croatia

Minobu Kasai
Department of Biology, Faculty of Agriculture and Life Science, Hirosaki University, Japan

Guillermo Noriega, Carla Zilli, Diego Santa Cruz, Ethel Caggiano, Manuel López Lecube, María Tomaro and Karina Balestrasse
University of Buenos Aires, Consejo Nacional de Investigaciones Científicas y Técnicas, Argentina

Gustavo B. Fanaro and Anna Lucia C. H. Villavicencio
Instituto de Pesquisas Energéticas e Nucleares (IPEN), Brazil

Qing Yu
Eastern Cereal and Oilseed Research Centre, Agriculture and Agric-Food Canada, Ottawa, ON, Canada

Printed in the USA
CPSIA information can be obtained
at www.ICGtesting.com
JSHW011414221024
72173JS00004B/540